TJ
2263
. P63
1998

HEAT EXCHANGERS:
A Practical Approach to Mechanical Construction, Design, and Calculations

M. Podhorsky
H. Krips
Deutsche Babcock
Balcke-Dürr, Düsseldorf, Germany

English Edition Editors:
William Begell, *U.S.A.*
Mike Morris, *U.K.*

with a Foreword by **André Bazergui,** *Canada*

Begell House, Inc.
New York • Wallingford, U.K.

an affiliate of Byelocorp Scientific, Inc. and Supco, s.r.l.
New York - Milan, Italy

HEAT EXCHANGERS: A Practical Approach to Mechanical Construction, Design, and Calculations

Copyright © 1998 by Begell House, Inc.   All rights reserved.

Printed in the United States of America.

Except as permitted under the United States Copyright Act of 1976, no part of this publication may be reproduced or distributed in any form or by any means, or stored in a data base or retrieval system, without the prior written permission of the publisher.

Direct all inquiries to Begell House, Inc., 79 Madison Avenue, New York, NY 10016.

Library of Congress Cataloging-in-Publication Data

Podhorsky, M.
    [Warmetauscher.   English]
    Heat exchangers : a practical approach to mechanical construction, design, and calculations / M. Podhorsky, H. Krips ; English edition editors, William Begell, Mike Morris.
        p.   cm.
    Includes bibliographical references.
    ISBN 1-56700-117-3
    1. Heat exchangers—Design and construction.   I. Krips, H.
II.   Title.
TJ263.P63   1998   98-18447
621.402'5—dc21   CIP

HEAT EXCHANGERS

TECHNOLOGY FROM A GOOD SOURCE

The BDAG Group company, Balcke-Dürr GmbH, is one of the leading international suppliers of power engineering systems and components for power stations, waste

incineration plants as well as for the chemical and petrochemical sectors. It supplies heat exchangers and cooling towers, energy-saving systems, decentralized power generation systems and turnkey chemical plants. More than one thousand highly qualified engineers and design draughtsmen are on hand to provide comprehensive expertise and innovative problem solutions. Worldwide, of course.

BALCKE-DÜRR GMBH D-40882 Ratingen Telephone ++49 2102 / 855-0

**COMPONENTS
SYSTEMS
PLANTS
SERVICES**

DEUTSCHE BABCOCK

BALCKE-DÜRR

BDAG GROUP

CONTENTS

Foreword vii
Preface ix
Introduction xi

1 Heat Exchangers for Power Stations 1
   1.1 Heat Exchanger Tubes 1
   1.1.1 Erosion and Corrosion-Erosion 2
   1.1.2 Droplet Impact 6
   1.1.3 Corrosion 8
   1.1.4 Crevice Corrosion and Contact Corrosion 11
   1.1.5 Stress Corrosion Cracking 14
   1.1.6 Vibration Damage 14
   1.2 Pressure Shell 14
   1.2.1 Heater Water Box 15
   1.2.2 Heater Shell 19
   1.2.3 Venting of the Steam Space 19
   1.2.4 Drain Cooler 21
   1.3 Vessel Shell 24
   1.3.1 Feedwater Tank and Deaerator 26
   1.4 Multistage Heater 29
   1.5 The Heat Exchanger for Process Heat Recovery 31
   1.5.1 U-tube Steam Generator 31
   1.5.2 Straight-tube Steam Generator 36
        Bibliography 49

2 Calculation of Structural Stresses Using the Force Method 51
   2.1 Introduction 51
   2.2 Application of the Force Method in the Construction of Heat Transfer Equipment 52
   2.3 Flat Heads 52
   2.4 Dished Heads 54
   2.5 Spherical Shells 60
   2.6 Cylindrical Shells 62
   2.7 Conical Shells 64
   2.8 Tubesheet 70
   2.9 Rings 73
   2.10 Combination of Individual Elements 75
        Bibliography 78

3	Calculation of Tubesheets		79
	3.1	Introduction	79
	3.2	Design of the Tubesheet According to Standards	79
	3.3	How to Calculate the Perforated Section	80
	3.4	Example of Tubesheet Analysis	81
		Bibliography	84
4	Design of Flanges for Pressure Vessels		87
	4.1	Introduction	87
	4.2	Bolt Design	88
	4.3	Selection of Gaskets	90
	4.4	Dimensioning Flanges According to Codes	92
	4.5	Dimensioning of Flanges Using the Deformation Calculation	97
	4.6	Structure of the Restraint Diagram	114
	4.7	Causes of Possible Flange Leaks	126
		Bibliography	129
5	Methods of Fastening Tubes in Tubesheets and Headers		131
	5.1	Introduction	131
	5.2	Welding of Tubes	132
	5.3	Roller Expansion of Tubes	134
	5.4	Hydraulic Expansion of Tubes	143
	5.5	Fastening of Tubes by Explosion	175
		Bibliography	181
	Appendix		183
	Additional Bibliography		191
	Topical References		213
	List of Standards and Programs		215
	List of Materials		217

FOREWORD

I have been aware of and quite impressed by the work of Dr. M. Podhorsky and his co-author, H. Krips, for a number of years, in particular for their exceptional contribution to the development of the hydraulic expansion process for the manufacturing of tube-to-tubesheet joints. I was therefore delighted and honored to have been invited by Dr. Podhorsky to write a foreword for the English translation of his book on Heat Exchangers.

This book offers an interesting overview of procedures for designing heat exchanger pressure vessels; although the approach is based mainly on the German Code standards, it regularly refers to and provides comparisons with the ASME Code.

The authors draw extensively from their wide industrial experience combined with a thorough understanding of the theory underlying the code procedures. Their approach is systematic and easy to follow and they use numerous illustrations to make their point. They cover the field with just necessary details and leave it to the reader to complement his or her information from the available design codes.

Chapter 1 provides an excellent description of the many types of heat exchangers found in the field, their applications, and the many practical problems that could be encountered (in particular those related to corrosion and vibration damage). It is an excellent "entrée en matière" for the experienced and inexperienced designers alike.

Detailed design procedures are given in Chapters 2 and 3. While Chapter 2 provides details on the discontinuity analysis approach for computing the stresses in the various pressure vessel components, Chapter 3 looks particularly at the design of tubesheets.

I was particularly interested in the authors' handling of the gasketed bolted flanged joint in Chapter 4. They are among the few who have tackled the problem in a complete manner, i.e., as a complex mechanical assembly whose purpose is to operate satisfactorily, not only in terms of pressure integrity but also in terms of leak tightness. It is, of course, beyond the scope of this book to cover in detail the extensive research work—more specifically that sponsored by the Pressure Vessel Research Council—that has recently been carried out of gasket evaluation alone. But it is interesting that their approach systematically takes into account the gasket as a full fledged mechanical element in the analysis. The fact that the ASME code is considering the introduction of new gasket factors and a new design procedure for gasketed flanged joints is an indication that the authors were pioneers in their approach.

Chapter 5 on Methods of Fastening Tubes to Tubesheets and Headers is, by itself, a justification to buy this book. It offers an excellent overview of the various methods available for securing a tight joint between the tube and the tubesheet. The authors cover all the methods in detail including, of course, the hydraulic expansion process which they pioneered. Having personally carried out research on the residual stresses generated in heat exchanger tubes by the various expansion processes discussed in the book, I can only but concur that the hydraulic process is the one that generates the least unwanted residual stresses while ensuring a systematic and uniform tube-tubesheet joint.

This book, first published in German, was long overdue in its English version. Hopefully it will contribute to a better understanding of two design philosophies which appear to be different but which are in fact basically quite similar. I wish the authors the best of luck and encourage them to pursue their excellent work.

André Bazergui, Ph.D., Eng.
Professor of Mechanical Engineering
Director General, École Polytechnique de Montréal
Montréal, Canada

PREFACE

Heat transfer, mass transfer, and storage are the three most significant elements in pressure vessel design. The sequence in which these are indicated here also reflects the significance in which the respective elements need to be considered.

The heat exchanger as a surface heat exchanger is not only the most frequent but often also the most sophisticated component in a plant. The problems connected with this component will be considered in detail. The procedure diagram below illustrates in rough stages the effects to be taken into account when designing the vessel; it is applicable for all vessels.

Standards such as ASME, BS, AD, TRD, etc., set out the minimum requirements for pressure vessel construction. If the conditions specified in the standards are complied with, then the vessel is designed appropriately for its operating conditions. This is, however, where the function of the standards ends. It is then the task of the design engineer to ensure that a component designed on this basis

- optimally fulfills its process function,
- has a high degree of operating availability,
- can be manufactured economically.

This book is intended to provide assistance in solving the questions that arise in designing a vessel. It is based on the extensive experience gained by the authors during their many years with Balcke-Dürr GmbH. The authors wish to thank the executive board of Balcke-Dürr GmbH for their constructive support and for consenting to the use of the drawings and photographs.

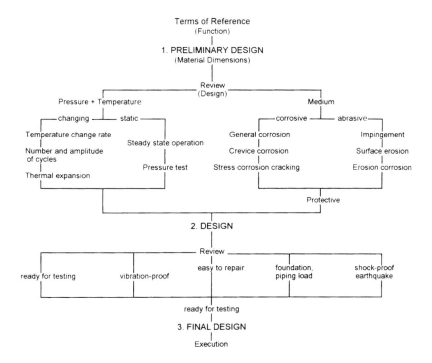

INTRODUCTION

The reliable operation of the individual elements of a system has taken on extremely great significance as a result of the rapid increase in both the capacity and complexity of the systems which involve vast amounts of capital expenditure. The heat exchanger is one of the components occurring most frequently in a system.

The book focuses on tube heat exchangers which contribute significantly to economic efficiency in the power station circuit. It covers sophisticated heat exchangers subjected to high operating loads from whose design and construction valuable information has been gathered. This information has been condensed here for the technically interested reader.

The first chapter deals with heat exchangers for power stations. It gives basic recommendations on the admissible loads, corrosion and erosion stresses, and on the main component assemblies. The basic process engineering rules and recommendations are also presented. Such information could therefore be valuable not only to young engineers but also to those with practical experience who might find ideas to improve a system through reading it.

The second chapter goes into the details of the fundamental elements which make up a heat exchanger or a pressure vessel. The deformation equations are given and explained so that the basic physical data of a stress analysis can be derived for a heat exchanger or pressure vessel subjected to internal pressure and thermal stress. The combination of the individual force method elements via the compatibility condition in the individual sections is specified to allow simple programming. The purpose of this chapter is to provide more basic information about the elementary components of the pressure vessel and to present the mathematical formulation in a clear manner.

The third chapter concentrates on some specific details relating to the design of a tubesheet. The tubesheet is a typical component of a heat exchanger which requires particular attention due to the geometrical inhomogenity in the section to be tubed.

The penultimate fourth chapter discusses the dimensioning of the various flange constructions in detail. Even today, a correctly designed flange joint is still unfortunately not a matter of course. The reason for this is the complexity of the joint which is subjected to the interaction of all the elements and also the fact that the design codes are in some cases simplified in a technically inappropriate manner. The gasket represents the additional unknown factor. It can have different sealing properties despite identical geometrical dimensions. The method of production and the material processing during production play a decisive role. Although the reliability of the entire plant during operation and during the start-up and shutdown procedures depends on the tightness of the heat exchanger connection, the problems relating to this connection are only seldom treated systematically and critically. This analysis covers both the flange design and the flange installation because these elements are directly interlinked. A correctly designed flange will leak if not correctly installed and vice versa. Attention is drawn to the fact that the deformation behavior of flanged covers differs fundamentally from the behavior described in the design codes and must therefore be treated separately.

The fifth and final chapter of the book also deals with a problem given little attention in the relevant literature and that is the problem of reliably fastening the tubes in the tubesheets. The authors have carried out very extensive work in this field. The invention of the process to hydraulically expand tubes into tubesheets and its introduction throughout the world are proof of this. The quality of the tube/tubesheet joint frequently determines the operating reliability of expensive and sophisticated process systems. Therefore it should be given due attention. All currently used methods of fastening tubes in tubesheets are set out and the advantages and disadvantages of each discussed.

The book was purposely restricted to those topics which can offer the reader new facts, the latest information gained through practical experience, or new solutions.

1
HEAT EXCHANGERS FOR POWER STATIONS

Power consumption and economic efficiency have led to a continual increase in the output of power generating units. Mass flow increased to the same degree as the output and consequently it became necessary to adapt the size of the components accordingly. This increase in size, however, is not just a question of adapting the geometry, because when certain limits are exceeded it becomes necessary to set new criteria. The extent to which such criteria can be incorporated in the design of heat exchangers by increasing their size is, however, limited. It then becomes necessary to split the component into two or several parallel units.

There is one heat transfer component element whose dimensions cannot be adapted by increasing its size: the heat exchanger tube, the outer diameter of which seldom exceeds 16 to 25 mm in heater construction. This restriction with respect to the tube diameter narrowed the scope for modifications in length in order to avoid high pressure drop or eroding velocity. Therefore, to achieve a higher output it is the number of tubes and not their size that has to be changed.

The number of tubes making up a heat exchanger or several parallel units of a thermal stage indirectly indicates the output of the plant. This number increases virtually in proportion to the increase in the size of the plant. However, the possibility of damage to the tubes and to their joints also increases in the same proportion. In the 1950's the average output of a power plant was 50 MW whereas currently the mass flow rate in a single heat exchanger corresponds to 700 MW and the number of tubes is 14 times as high. Therefore, assuming that the current components have the same operating life as the ones used previously, the damage safety margin for the tubing must be 14 times higher. This safety margin needs to be even higher because the increased costs incurred through a loss in running time of large industrial plants must be taken into account. It is therefore obvious that the measures to increase operating time should concentrate mainly on testing and protection of heat exchanger tubes. It goes without saying that the other pressure and structural parts have to meet specific safety requirements.

Proper design and material selection and good engineering produce high quality components only if accompanied by a stringent quality control system and if the manufacturing of the parts is monitored by appropriate testing.

Quality control and testing ensure economic efficiency. They help to guarantee running time and to reduce the manufacturing costs. Quality control personnel must, however, also be involved right from the design phase so that the testing required during manufacturing can be restricted to a minimum.

1.1 HEAT EXCHANGER TUBES

As a rule, tubes are made of ST 35.8 I, ST 35.8 III or 15 Mo 3*, depending on pressure and temperature. The tubes in low pressure heaters are joined to the tubesheet by rolling or by

---

*For material equivalence please see List of Materials at the end of the book.

hydraulic expansion using the **HY**draulic **T**ube **EX**pansion (HYTEX) process. The tubes in high pressure (HP) heaters are welded to the headers or the tubesheets and also hydraulically pressed or rolled in the tubesheets.

Conventional tube materials have proved to be suitable provided that the heat exchanger design takes into account the effects of pressure, temperature and medium involved. When designing a component, precautions must be taken to prevent: erosion, corrosion-erosion, droplet impingement, corrosion and vibration.

1.1.1 Erosion and Corrosion-Erosion

Erosion and corrosion-erosion are the main cause of damage to exchangers. These can be avoided by taking appropriate precautions.

On the water side (inside of tubes) the flow rate is kept within specific limits to avoid erosive vortices in the inlet flow (Fig. 1.1). The flow rate in the C-steel tubes is generally 1.8 m/s and lower; this rate may only be increased to 2 m/s if the tube openings have been appropriately designed. The maximum flow- rate for stainless steel tubes is 2.5 m/s and in this case plant-specific criteria such as pressure drop are the decisive factors. Lower flow rates can also cause corrosion-erosion in C-steel tubes if the pH value is below 9. Particular attention must be paid to flow conditions with these modes of operation.

It is not the rate of the mass flow alone which causes damage; the geometry of the water box and the header also plays a role. When determining the number of tubes required for a rate of 1.8 to 2 m/s, it is assumed that the flow rate through each of the tubes is identical. Apart from some tolerable variations which occur due to the difference in length of the internal and external U-tube, this is normally the case. This uniformity is frequently disrupted by water boxes which are too small, because the water inlet flow does not have an adequate distance to develop. Inlet nozzles arranged laterally and immediately upstream of the tubesheet pose a danger for the tube inlets as they entrain vortices and disrupt distribution. It is particularly important to take this into account for HP heaters operating in the temperature range above 150°C. It may be necessary to provide a flow rectifier here in addition to rounding off the tube inlet. (Fig. 1.2).

The streamlined openings of the rectifier are aligned with the tube openings. Their diameter corresponds to the inside tube diameter. The ratio of sheet thickness to opening

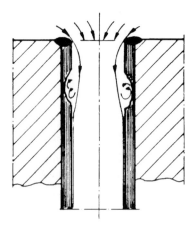

Fig. 1.1 Erosion in the tube inlet.

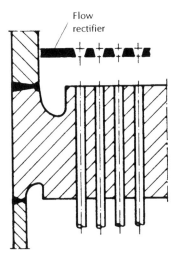

Fig. 1.2 Flow rectifier.

diameter should be ≥ 1. The distance between the rectifier plate and the tubesheet should be at least twice the tube diameter so that vortices at the rectifier outlet cannot reach the tube inlet.

On the steam side (outside tubes) the damage caused to the bundle by erosion and corrosion-erosion is more diverse. The flow rate is always the decisive factor. The inlet flow, the distribution in the vessel and the approach flow to the heating tube surface should be checked. In addition, the conditions in the water vapor loop should be monitored during operation. The first measure is to reduce the flow rate; this, however, is restricted by economic considerations. The risk of erosion can be overcome to a significant extent by adapting the construction. The first step is to take into account the state of the steam at the steam inlet. In a conventional power station the high pressure steam is superheated and the low pressure steam from the lower extraction point is wet. In a nuclear power station with a pressurised-water or boiling-water reactor, the heating steam entrains larger quantities of water into all stages.

The diagram in Fig. 1.3 shows admissible steam inlet flow rates obtained from practical applications. It is assumed here that the steam is dry or superheated and that the tube bundle is provided with impact protection. The impact protection of the tubes takes on greater significance, should there be specific grounds for higher inlet flow rates, e.g., restricted steam piping diameter due to lack of space. The same applies in the case of wet steam which always requires lower flow rates. The extent to which protection measures are required, depends on the individual droplet spectrum.

In the steam inlet section the tubes are provided with impact grids to protect them from the kinetic energy of the stream. The grids have proved more effective than previously used impact plates for several reasons. It is important that the velocity energy of the entrained water is reduced and as much water as possible is separated from the steam before it reaches the tubes.

The impact grid consists either of expediently arranged shaped sections, round steel bars or impact tubes. In U-tube vessels, the outer tube rows of the bundle can act as impact protection, provided that the tubes in these layers are made of stainless steel. In this case, the impact protection would also be involved in the heat transfer. The impact elements are

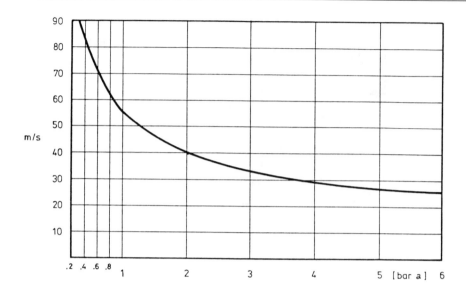

Fig. 1.3 Standard flow rate for dry steam in steam inlet nozzles.

arranged so that no streams of water are deflected in the direction of the heater shell where they could cause damage. Fig. 1.5 shows a V-shaped steam lane with an impact grid made of angle elements. The arm of the angle is arranged in such a way that the direct path to the tubes is blocked.

When an eccentric bundle arrangement with circular cross-section is selected, an impact grid made of tubes is preferable. The impact tubes disperse the impacting droplets and therefore prevent an inlet flow throw-back which would damage the shell. The tubes of an eccentric bundle which are not protected by impact grids should be covered by a protective shell which extends over the remaining circumference, Fig. 1.6. An ejecting flow can occur over the exposed tubes at the inlet to the crescent-shaped lateral sections if the inlet flow section is not covered in this manner. That means that the moisture from the steam which has already condensed is drawn in and flung against the peripheral tubes. This secondary moisture can also cause erosion even in the case of superheated steam.

The axial flow and the flow into the bundle also represent a risk factor as far as corrosion-erosion is concerned. The steam lanes and the inlet flow surface of the bundle periphery must be adequately dimensioned. Standard flow rates based on experience are indicated in Fig. 1.4.

The term "thermal symmetry" was coined for an economic steam lane shape. It is only possible to use the most economic shell diameter when the bundle arrangement is thermally symmetrical (Fig. 1.7). Thermal symmetry can be most easily achieved with a circular cross-section as shown in construction 2 B in Fig. 1.7. There are, however, important points in favour of the V-lane, e.g., the entire length of the bundle can be inspected. With V-lanes the lefthand and righthand tube field sections are inclined, refer to A 2 in Fig. 1.7, or the tube field is distorted as shown in Fig. 1.8. The latter method is more sophisticated but the construction and the bending plan for the U-tubes are more complicated.

The steam condenses on the tubes and is consequently drawn into the vessel, its flow direction is then determined in the heater. It is, therefore, necessary to consider the different condensation capacities of the individual bundle sections. The size of the steam lane and the free space for the longitudinal flow cannot be determined until the actual distribution is known. The graph in Fig. 1.9 shows how the distribution of the axial steam flow is determined.

Curve A in the graph in Fig. 1.9 indicates the condensation output of the bundle as a function of the unit of length. It is based on the output of the cold and the warm arms of the U-tube, a and a'. Curve B indicates the total condensates. Here again the values for the cold arm b and the hot arm b' are combined in B.

The correctly determined side-stream rates and the thermal symmetry based on the admissible flow rates are required to optimise the shell diameter. If the flow rate is too low, the vessel will be unnecessarily large; an excessively high steam rate causes corrosion-erosion at all obstacles in the flow such as edges of supporting walls or anchor rings. Under certain circumstances secondary moisture (aspirated condensate) increases this corrosion-erosion.

The effect of the flow rate depends essentially on the gas content of the steam and its pH-value. If the pH-value is reduced , increased O_2 and CO_2 contents can disturb the formation of a protective film even deep inside the bundle where the flow rate is low if the pH value decreases. In particular this affects the parts of the tubes located in the supporting wall sections. Subcooling in the gap between tube and supporting wall causes aggressive concentrations of gas due to the fact that the gas is more soluble; therefore this circular section is susceptible to greater corrosion-erosion as shown in Fig. 1.10. For this reason the supports in the condensation section are made in a grid-type construction. Subcooling is

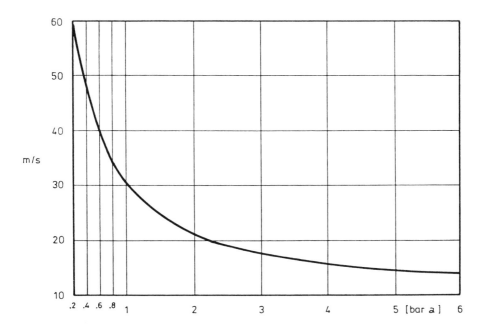

Fig. 1.4 Standard flow rate in steam lanes.

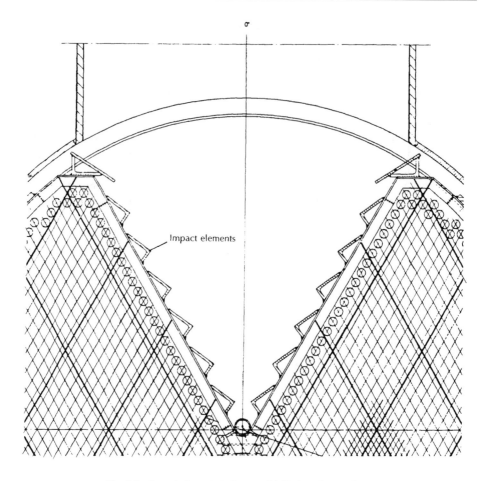

Fig. 1.5 Impact elements in heater with V-shaped steam lanes.

virtually eliminated and the longitudinal velocity reduced as a result of the linear contact between the grid rod and the tube. (Fig. 1.11)

1.1.2 Droplet Impact

Droplet impact erosion is caused as a result of the kinetic energy of the droplets being transformed into deformation energy on the tube surface. In appearance, this erosion resembles the surface of sand paper. The energy of the droplets is destroyed by the plastic deformation of the sensitive sections of the structure. This constant process causes the displaced material to harden and the corroded particles to splinter off. At first it is the ductile elements of the structure that are affected whilst isolated hard spots resist (emery); as erosion progresses the harder elements of the structure are also released.

Droplet impact erosion can occur with wet heating steam and when expanding condensate flows into the vessel. The effect is determined by the size of the droplets; the mass of the droplets increases with the droplet diameter to the power of three. The velocity conveyed to the droplets by the steam increases the droplet energy by the second power. The reduction in enthalpy produces a droplet size in the microrange which would have no technical significance at the velocities which may occur in piping and vessels. Inevitable disruptions of the flow on the way to the heater do, however, cause agglomeration and the resultant droplet spectrum is therefore not predictable. In every case precautions must be taken to avoid the risk which this entails for the heat exchanger tubes in the saturated and wet steam sections.

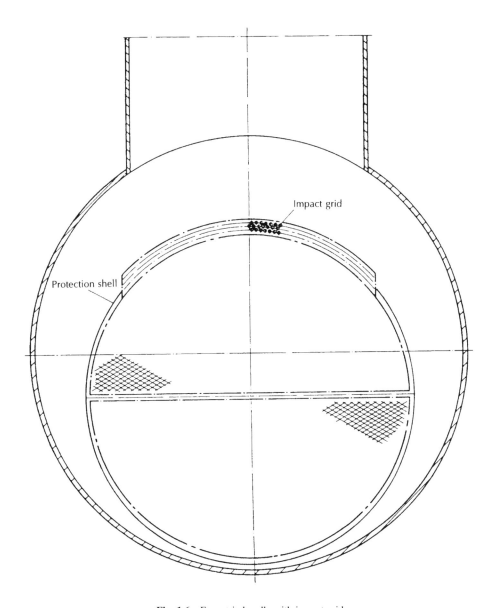

Fig. 1.6 Eccentric bundle with impact grid.

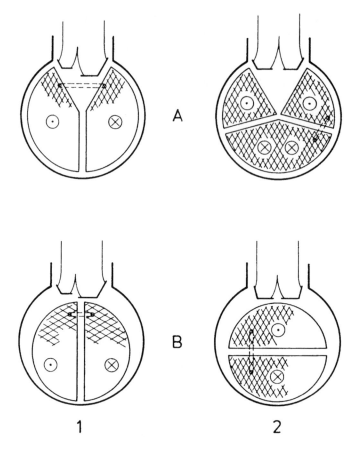

Fig. 1.7 U- tube arrangement. A_1–B_1, non-uniform bundle approach flow; A_2–B_2, thermal symmetry: X, cold branch; •, hot branch.

This type of damage can be prevented by protecting the tubes against the inflowing steam with impact grids and baffle plates and also by adjusting the velocity appropriately. The standard velocities indicated in the graphs in Figs. 1.3 and 1.4 should not be exceeded, if possible.

1.1.3 Corrosion

If the flow itself has no effect and if corrosion elements inherent in the material can be excluded in the case of steel tubes, the only remaining possible cause of corrosion is the low resistance of the material towards the medium. Inadequate heating steam or condensate quality during operation and air intakes during outages give rise to damage on the heating surfaces [1].

As a rule, the form of the corroded areas clearly indicates the point of entry of the aggressive medium. In a vacuum tank, that point is not necessarily in the steam inlet section (air intake). As long as the corrosion remains in the form of a surface film, the heat exchangers tubes have not been seriously damaged. In due course, however, corrosion nuclei form beneath this surface film, which consists primarily of ferric hydroxide FeO (OH), and these then lead to voluminous patches of rust with dangerously deep corrosion troughs beneath them.

The construction cannot be adapted to provide protection against this as the substances causing corrosion are either entrained in the steam or enter the system with the air. The solution is, however, to control the chemical composition of the water in the loop and provide appropriate protection during outages. Stainless steel tubing is selected only under exceptional circumstances and this is then restricted to the low pressure (LP) section. Peak-

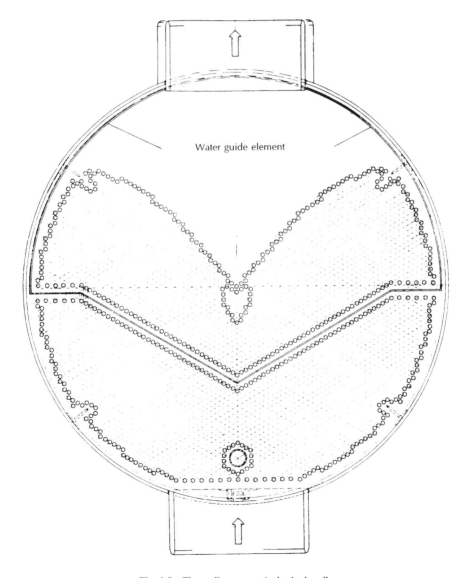

Fig. 1.8 Thermally symmetrical tube bundle.

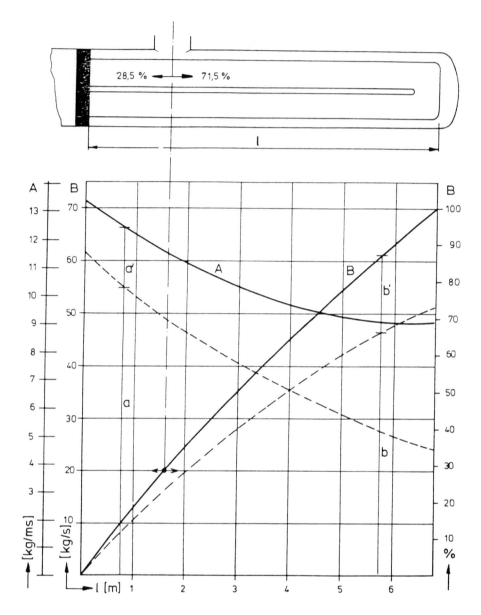

Fig. 1.9 Mass flow in the heater. A, length-related condensation output; B, cumulative condensation output.

load power stations, which have frequent outages related to output demand, require short-term preservation measures which are difficult to implement effectively. In this case it is advisable to use stainless steel tubes.

Little benefit is gained from preserving the components comprising C-steel tubes with nitrogen when the bundle is damp. The electrolyte formed by the moisture produces the oxygen required for the corrosion process at the corrosion front. It is important that the tube bundle is dry and that during the outage the temperature does not drop below the dewpoint of the gas cushion, whereby the dewpoint temperature depends on the moisture content of the gas. If the temperature of the specific bundle remains always above the dewpoint, then any gas can fulfill the task of the nitrogen, even dry air.

In the case of peak-load power stations with regular short-term outages, the LP heater bundles cannot be dried quickly enough with hot air and therefore the use of C-steel tubes always poses problems. A more effective method of providing protection is to circulate the hot contents of the feedwater tank. The hot water which is always above 100°C is tapped from the feedwater tank and returned via the LP heater. In this way the heating surfaces are always maintained above the dewpoint. In order to heat the heating surfaces in this way, however, it is necessary to provide a pump and an extensive automatic control system to protect the heater from temperature shocks; therefore, in the long-term stainless steel tubes are regarded as a more economic form of corrosion protection. The same corrosion phenomena occur if ventilation is not adequate during operation. The non-condensable gases including oxygen become concentrated to saturation point and induce corrosion when the protective film is inadequate. A heater is properly vented only if the entrained gases are extracted at the ends of the condensation flow. The gas can be conveyed to the vent points by internals (refer to Venting 1.2.3). The gas can only be discharged with the optimum amount of entrained steam, i.e., the lowest possible amount, when the condensation flow is properly adjusted.

Steam baffles are critical as far as corrosion is concerned because final condensation points can form behind them and consequently produce also gas accumulations. This is another important reason for constructing the baffle protection elements as steam-permeable baffle grids.

1.1.4 Crevice Corrosion and Contact Corrosion

As far as heat transfer equipment is concerned, this type of corrosion which occurs predominantly localised in crevices can be attributed mainly to the concentration of substances. These develop because the crevice is not well flushed and the subcooled condensate present there increases the solubility of the corrosive gases thus enabling them to become more concentrated.

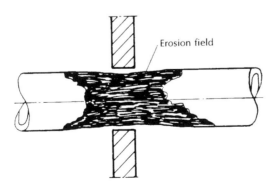

Fig. 1.10 Corrosion-erosion (erosion field).

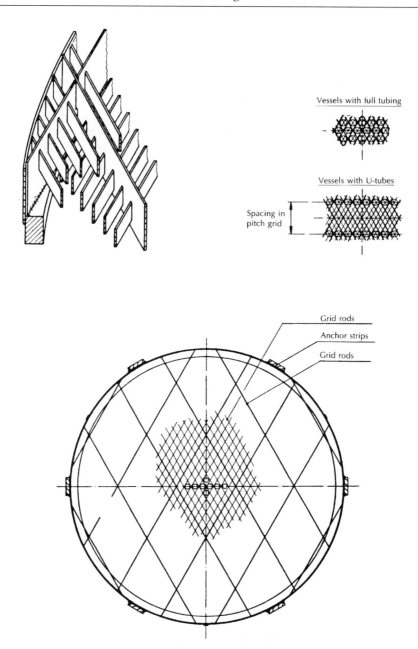

Fig. 1.11 Tube support grids, inserted type, 60° pitch.

Another form of concentration in the crevice occurs during evaporation when the warmer medium flows through the tubes. All crevices between the outer surface of the tube and the material in contact with it cause the aggressive constituents in the water to become concentrated due to the lack of flushing.

The tube/tubesheet joint and passage through the supporting walls are affected by crevice corrosion. There should be no other contact points between the tubes and the internals.

If the tubes are rolled into the tubesheets there is a gap on the rear side of the tubesheet. Considerable forces are required during the rolling process to plasticize the tube wall. A safety clearance of approximately 5 mm is provided between the rolled-in section and the rear surface of the tubesheet to prevent the tubes from being indented or even sheared by these forces behind the tubesheet. The subsequent gap between the outer wall of the tube and the inner wall of the borehole can lead to crevice corrosion. Stainless steel tubes are at particular risk because in the transition between the rolled and non-rolled sections of the tube, i.e., in the gap, very high tensile stresses are transferred to the outer surface of the tube [2].

It is possible to secure the tubes with no gap and minimal stress by hydraulically expanding the tube ends into the tubesheet (HYTEX process).[1] The expansion distance then extends to the end of the borehole on the rear face of the tubesheet. This is a good and economic way of excluding crevice corrosion on the tubesheet. Another, more complicated solution is to weld the tube ends to the rear face of the tubesheet without a gap.

The great significance placed on the tube/tubesheet joint and the wide range of problems which can occur in this connection are described in Chapter 5, which also gives details of other possible types of joints.

The situation with respect to supporting walls is similar to that of the supporting walls at the rear of the tubesheet. The gap between the tube and the supporting wall borehole becomes filled with saturated, subcooled condensate which results in typical dual ring corrosion when flushing is not adequate. (Fig. 1.12). This damage can be avoided by using a grid-type tube support. The crevice is restricted to a short line of contact at the grid bars. This section can be flushed well.

Denting, a particular type of damage caused to the tube by the crevice in the supporting walls, can also be avoided by using the grid-type support. Denting occurs in evaporators or steam generators in which a hot primary medium flows through the tubes to evaporate water in the shell section. Solids contained in the water (e.g., phosphates) become concentrated in the annular gap between the tube and the support wall borehole,

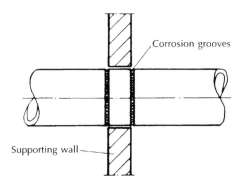

Fig. 1.12 Dual ring corrosion.

---

[1]HYTEX® - Expansion process developed by Balcke-Dürr AG.

ultimately filling this gap completely and causing a buildup of pressure. Salts and corrosion products can cause a reduction in the tube cross-section and thus bring about the denting phenomenon [3]. Contact points between internals should also be minimised as far as possible, in particular at the condensation limits where not only flow is absent but where the gas partial pressure is at its highest.

1.1.5 Stress Corrosion Cracking

Should the mode of operation require the use of stainless steel tubes, then stress corrosion cracking should also be taken into account. In this case it is the transition between the rolled and non-rolled section which needs attention. High stresses occur in this transition section due to mechanical deformation leaving considerable residual stresses in this zone. This stressed section is particularly at risk when the hot medium in the heat exchanger involves flows through the tubes (steam generator). Corrosive constituents accumulate in the gap which inevitably forms when the rolling process is applied; consequently crevice corrosion and to a greater extent stress corrosion cracking can occur [5].

It is not only a matter of closing the gap but of keeping the residual stress low. In this case the previously mentioned HYTEX process is particularly suitable in two respects; firstly it closes the gap and secondly hydraulic deformation causes the least residual stress.

1.1.6 Vibration Damage

Based on current knowledge about the vibration excitation of tubes [4, 5] damage caused by vibrations can be completely avoided provided that the proper measures are implemented. It is obvious that

1. there must be adequate clearance between the excitation frequency and the natural frequency;
2. the tube supports have a cushioning effect with, for example, wide grid bars resting on the tube wall; and
3. the support clearances are non-resonating.

Particular attention must be paid to the tube bends. In this case, they must be provided with support in the bend section unless they are in a condensation section and are therefore in an area with a low flow rate.

1.2 PRESSURE SHELL

It is advisable to restrict the different types of material used and to make the selection according to strength properties, purity of the melt with respect to sulphur and phosphorous and to good working properties, in particular weldability. Bearing these aspects in mind the materials currently selected for the heating section and the feedwater tank are WStE 255, WStE 355, 15MnNi63, 20MnMoNi55 and 15NiCuMoNb5 (WB36).

The diameter of the heater shell is determined by the dimensions of the tube bundle and the admissible flow rate for the bundle. Furthermore, the pressure shell should not have a complicated structure, i.e., geometrical discontinuities (significant changes in wall thickness, expansion restrictions, force line disruptions, etc.) must be avoided as far as possible. This fact becomes more significant when the load on the wall is greater and the stress level is consequently higher.

The pressure shells are designed in accordance with specific standards. Discontinuities which arise, for example, due to unavoidable wall thickness changes or tubesheet/shell welds are not, however, taken into account. The rotational symmetry of the pressure shell can be easily checked using the force method (see Chapter 2).

1.2.1 Heater Water Box

The flanged water box is selected mainly for vessels with smaller diameters so that the tube inlets are more easily accessible. This type of water box may also be required for vessels with larger diameters if the tubesheet or its cladding consist of non-weldable material, e.g., process water or plume coolers.

It is obvious that the flange connection is of such significance that a calculation taking into account all the effects is essential for proper design (refer to Chapter 4). Apart from the internal pressure and the temperature which the connection must withstand, the tightness will depend on the deformation occurring when the loads are absorbed. As a rule, the water box has a pass partition which relays the different temperatures of the medium flowing through to the appropriate flange sections. In heater systems in which the heaters can be bypassed, the water boxes of the upper heaters must be able to absorb a large range of temperatures, in some cases up to 100°C. Under these circumstances the design engineer should, if possible, change from a 2-pass to a 4-pass vessel in order to minimise the stresses at the flange joint resulting from temperature differences.

The partition wall ligament and the tubesheet should be brought into close contact by placing a gasket between them to ensure the tightness of the pass partition under all operating conditions. If the sealing areas are permeable, the gasket material is soon washed away. The bypass which then forms not only reduces the output of the heater but also erodes the tubesheet and the ligament until these are no longer fit for use. HP heaters are particularly at risk in this respect because of the high water temperature.

The pressure in the water box of a heater is always higher than the pressure in the shell section. The tubesheet deforms as a function of the differential pressure between the water box and the shell section so that the gasket is subject to less pressure at the centre of the tubesheet than at the edge. At the same level of stress in the tubesheet, the pressure at the centre of the tubesheet and that at the edge differ more greatly as the diameter of the vessel increases.

As the gaskets have only limited resilience, the tubesheet thickness should not be calculated according to strength criteria but according to the deflection admissible for the gasket. For this reason a type of material with the lowest possible strength should be selected for the tubesheet, taking into account the mechanical and hydraulic tube fastening, the welding, and heat treatment techniques used.

The sensitive element of the flange construction has been eliminated by using the more economic welded water box construction whenever technically possible. There are, however, other problems to resolve.

Normally, the welded water box has two critical points: firstly the water box shell/tubesheet joint and secondly the pass partition for the water flow.

The transition from the tubesheet to the water box shell is provided with sufficiently large radii to allow peak stresses which inevitably occur in this groove section to be relieved and the fatigue damage to be reduced (Fig. 1.13). Even in the case of a well-formed

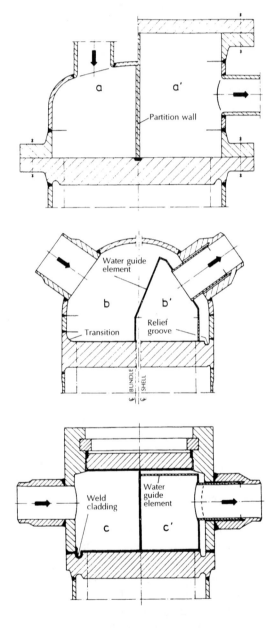

Fig. 1.13 Water box constructions. Flanged: "a" with hood, "a'" with cover; welded: "b" with forged transition, "b'" with transition in the relief groove; self-sealing: "c" clad transition, "c'" transition as b'.

transition, this point is very sensitive to changes in temperature. The transitions should therefore be ground so that machining furrows do not become the starting point for incipient cracks.

When a protective film has built up, exceptional and frequent changes in stress levels cause cracks in the magnetite and hematite layers which are less ductile than the base material. This causes the development of cracks resulting in considerable damage due to the Schikorr reaction [1]. In order to prevent initial cracks in such cases, the high stress zone in the groove sections is weld-clad with an austenitic material.

The pass partition prevents expansion if it is constructed as a rigid separation wall welded in place. This construction is justifiable in the LP-section. A separate water box is always installed in the HP section to guide the water. The need to install such a guide water box depends on the extent of expansion, i.e., on the strength of the material and the wall thickness of the water box.

Material with higher strength properties is required in the HP section. It is only appropriate, however, to use such material if its strength and therefore its expansion properties are utilised to the full. If a rigid separation wall is used, then the joint weld has to absorb this greater expansion and this would cause high stress peaks in the transition section to the water box shell. A rapid drop in temperature would force this stress level up significantly because the temperature of the thin partition wall in contact with the flow on both sides leads the temperature of the thick water box. The multiaxial stress cycle would considerably limit the service life of this component. Therefore a separate water box is required here.

A guide water box must also be installed if the circumferential weld between the tubesheet and the water box shell is to be examined. If a rigid separation wall is used, the weld section cannot be subjected to an assessable volumetric examination.

The same criteria apply to the water box with the self-sealing closure. The particularly extreme difference in the wall thickness of the water box wall and the partition wall gives rise to cracks in the welded section of the pass partition. Heaters of this type have a restricted diameter and therefore parallel units are installed to provide the necessary output. In general, such heaters are constructed with a water box diameter of up to 1200 mm.

In industrial power stations an individual vessel with a high output is, for economic reasons, always given preference over several vessels with appropriately lower outputs connected in parallel for economic reasons. When HP heaters with welded or flanged water boxes are used, it is no longer expedient to conform to this due to the high pressures. The water box with self-sealing closure cannot be used here either because the water box diameter would no longer be of a manageable size. Therefore, the header must be selected as a special type of water box. (Fig. 1.14).

The diameter of the inlet and outlet headers is considerably smaller than that of the water boxes and therefore the pressure and the temperature can be controlled better. The heat exchanger tube connections should be arranged over the entire header area if possible so that a uniform strength reduction factor evens out the stress.

A good distribution of the header boreholes also has a favourable effect on the flow conditions. As is the case with plate heat exchangers, the aim here is to avoid erosion in the boreholes and the connecting nipples. Operating experience has shown that the water flow rates in the header inlet should not exceed the 3.5 m/s and it is essential that the edges of the holes are rounded (R > 5 mm) as this reduces the stress at the borehole rims.

18 **Heat Exchangers**

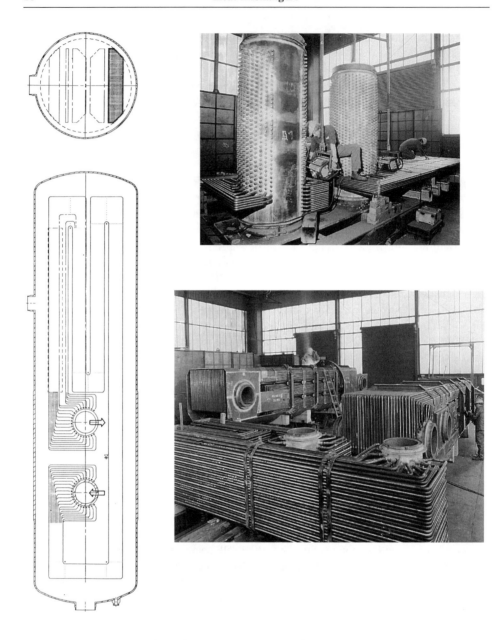

Fig. 1.14 Construction of headers for HP heaters.

The penetration section of the pressure shell and the thick-walled header is subjected to particular loads because the material mass in the weld section of this joint causes thermal stresses under temperature fluctuations.

The admissible temperature gradient must be determined here in the same way as for other constructions and indicated in the operating instructions so that the start-up and shutdown procedures can be regulated accordingly. High stress peaks are caused by the temperature drop brought about by unforeseeable shutdowns or automatic bypassing of the heaters. Such occurrences and the normal load fluctuations are estimated for the service life of the vessel, summarised in a loading spectrum and used as a basis for the fatigue analysis. Compliance with the results as per TRD 301 Attachment 1 often leads to unfavourable gradients which cannot be achieved in operation. A specific re-calculation is then required in order to determine whether the normal operational temperature gradients can be absorbed by the given component [6].

1.2.2 Heater Shell

There is a risk of damage to the shell both in the steam inlet section and also at the inlet due to the expanding secondary condensate. The steam nozzle for superheated steam is provided with a thermal protection tube in order to prevent great differences between condensation/shell temperature and superheating/ nozzle temperature. In the case of wet steam, baffle elements are provided to prevent the inflowing steam from being deflected against the shell. The same applies for the expanding condensate. In this case the nozzles are appropriately protected (stainless steel inserts, cladding, wear inserts) and the flash tanks lined to prevent erosion from occurring.

The secondary condensate flowing out of the flash section may still contain steam bubbles from the flashing process. It must be ensured that these bubbles are not entrained into the condensate outlet or the cooler inlet. The consequence could cause fault conditions at the secondary condensate pumps or cavitation on the cooler tubes. A weir or other suitable equipment prevent such damage.

1.2.3 Venting of the Steam Space

The heating steam will always contain traces of gas. These, however, have hardly any influence on the heat exchange in the first condensation zones. On the way to the final condensation point, the relative gas content rises as a result of the declining steam content. At the end of the condensation process there is then a steam/gas mixture with a gas partial pressure which brings about subcooling. The lower limit of this subcooling is determined by the temperature of the water flowing through the tubes in this section.

Subcooling and, therefore, the partial pressures remain constant for the final phase of the condensation process at the cold heat exchanger tube inlet. This applies even if the gas content is exceptionally low. At the other final condensation points (there are always several final points) subcooling is always determined by the temperature of the water which has reached the point in question. If the system is poorly vented and the constant supply of gas is not discharged, then the subcooled zones would become larger causing temperatures to decrease and subsequently an exchange of heat would no longer be possible.

In order to vent the system properly, the steam/gas mixture must be drawn off at all final condensation points. It is obvious that the amount of mixture which has to be extracted to remove the specific gas quantity from a tube section in which the water is already heated is greater than that to be extracted from sections with lower temperatures and, therefore, least of all from the cold tubesheet section.

Irrespective of the extraction point, the vent point must always be screened off so that only gas flowing from the bundle, i.e., subcooled gas, can reach it. It is essential to prevent the non-subcooled heating steam from taking the direct path to the vent system.

It must be made possible to adjust each individual vent connection and the collecting piping must be dimensioned to ensure that each connection can discharge its steam/gas mixture.

The aim is always to reduce the number of final condensation points. With HP heaters this aim can be achieved by providing the steam inlet directly behind the tubesheet. In the low pressure section the inlet must be located in the centre due to the large steam volumes and this inevitably leads to a greater number of vent points. Appropriately designed internals could also provide a good solution here. Fig. 1.15 illustrates an upright heater in which the problem of venting is resolved by providing an internal guide shell. The guide shell extends below the condensate level and thus directs the steam in the appropriate

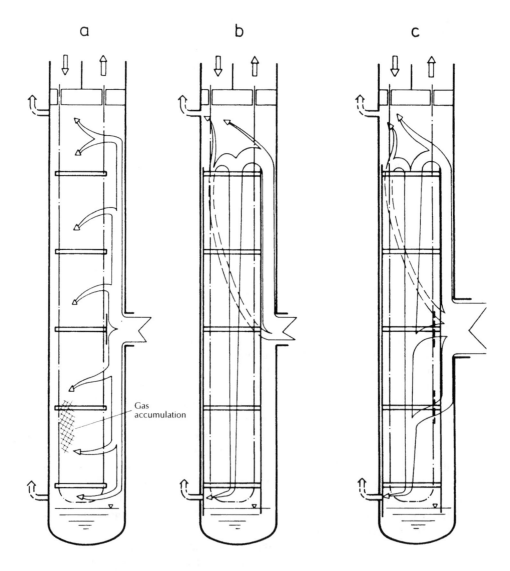

Fig. 1.15 Venting of upright heaters: a, poor; b + c, good venting conditions.

direction for optimal venting. When very large steam volumes are involved, additional openings are provided to enable the steam to enter the bundle. Such openings must, however, be restricted in order to maintain the downward flow.

1.2.4 Drain Cooler

It is advisable to provide the heating condensate cooler as a separate vessel. The heater and cooler continue to be combined for reasons of economy and space and this means that additional functional conditions have to be fulfilled.

The coolers in HP heaters can be installed either upright or horizontally. In LP heaters the outlet pressures are normally so low that geodetic height of the upright cooler cannot be overcome at low load due to prior flashing of the condensate. The cooler can therefore only fulfill its function in the LP lane in horizontal vessels (Fig. 1.16 and 1.17).

The tightness of the cooler shell is a fundamental requirement both for an upright and for a horizontal cooler fill. As the pressure in the cooler section is always lower than in the condensation section, a flaw in the cooler shell allows steam or flash-off condensate to penetrate the cooler. Any heating or flash-off team which has penetrated the shell first, collapses in the subcooled condensate and enlarges the flaw through cavitation. This allows the inflow to increase until finally the cooler tubes are affected by cavitation and the tube wall is eroded to the state of rupture within a short time.

As it is very difficult to repair weld damage on the cooler shell if it is located in the inaccessible section of the tube field, it is important to arrange all welds so that they are exposed when the bundle is removed. The inevitable concealed joints must be subjected to a stringent examination.

A further important requirement for the cooler is a well functioning and reliably serviced drain control for the heating condensate. Steam penetrates the cooler if the condensate level drops to the level of the cooler inlet because the controller is slow to react. The resultant condensate shocks subject the cooler shell to considerable strain.

The shocks can cause deformation in the less stable flat wall section with low stability or surface cracking in the peripheral section. It is therefore advisable to provide the upright cooler in the vertical heater with a cylindrical shell which has a greater resistance to condensate shocks (Fig. 1.16). The circular cooler shell is not used in the horizontal heater because a lateral inlet to the cooler affects the stability of the cylinder. A significantly larger heater diameter and therefore a more expensive vessel would be required in order to provide an axial inlet by submerging the cooler. For this reason an immersion tube renders the horizontal construction more reliable, above all when the control range in the heater shell is very low (Fig. 1.17). The immersion tube only safeguards against exceptional short fall below the control range. The control level should never be located in the immersion tube section.

In the condensation section of both the HP heater and the LP heater, the water level should never be below the lower cooler tubes because otherwise flash-off condensate, i.e., condensate containing steam would be conveyed to the cooler tubes and cause cavitation. In this case it is the cooler of the heater with the lowest pressure which is most at risk.

The minimal subcooling of the condensate at the cooler inlet acts as the reserve and protection against cavitation. The differing rates at which this reserve is used in the various pressure ranges is clearly illustrated in the water column equivalent for subcooling in the graph (Fig. 1.18). When subcooling is 1°C, a heater with a steam pressure of 8 bar a has

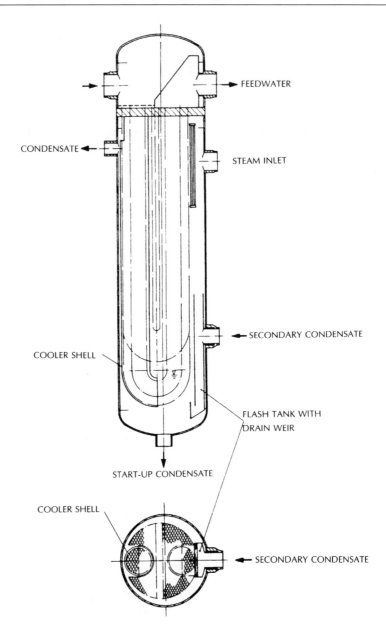

Fig. 1.16 Upright heater with built-in condensate cooler.

a reserve of approx. 2 m water column. A heater with a steam pressure of 1 bar a, however, only has a reserve of 0.4 m water column. The water column in the graph represents the sum of the pressure drop and the geodetic height and takes into account the temperature-related densities.

If venting is good and if the tubes in the condensation section are not submerged, which is only possible in a horizontal heater, then subcooling is minimal. It can be seen from the graph (Fig. 1.18) that in the pressure range of HP heaters, i.e., above the 8 bar taken into account, the slightest subcooling provides adequate reserve against cavitation. In the case of LP heaters, however, the reserve provided by subcooling is soon used up, especially if they are operated in the low load range.

Steam is released as a result of a fall below the boiling pressure if the subcooling reserve has been used up due to pressure drop or geodetic height. The cooler in the LP heater is again affected most as the cavitation intensity depends on the steam volume. The graph (Fig. 1.19) shows the boiling pressure as a function of the evaporation volume. For example, approximately 6 litres of steam are released per kg of condensate at a boiling pressure of 0.5 bar after a fall of 0.2 m below the water column. The same steam volume occurs at 3 but not until the fall below the water column is 4 m.

A cooler plate (Fig. 1.17) is required to separate the cooler from the condensation section in horizontal U-tube heaters. The tubes leave the cooler section by passing through the cooler plate and then they form the condensation surface. Steam flows in the direction of the cooler section via the annular gap between the plate borehole and the tube. Steam must be prevented from reaching the cooling section because cavitation would occur there. The gap must be as narrow as possible and the plate as thick as necessary. This measure should cause the steam flow entering the gap to condense and be cooled on the way through the plate. The most reliable way of excluding cavitation is to roll in or hydraulically expand the tubes to the wall of the boreholes in the cooler tubesheet. It is only possible to roll the tubes in to a limited depth. Hydraulic expansion (HYTEX) can be used without any restrictions.

Finally it should be noted with respect to drain coolers that they must always be arranged lower than the control level in the appropriate heater and as a separate vessel. If the condensate is drained via a loop, the coolers must always be located at the lowest point of the rising branch. An integrated cooler may not be used if there is a drain loop. Apart from the fact that this would mean that cooling is not achieved, there would be a risk of considerable condensate shocks.

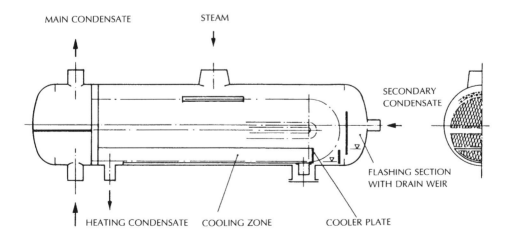

Fig. 1.17 Horizontal LP heater with integrated drain cooler.

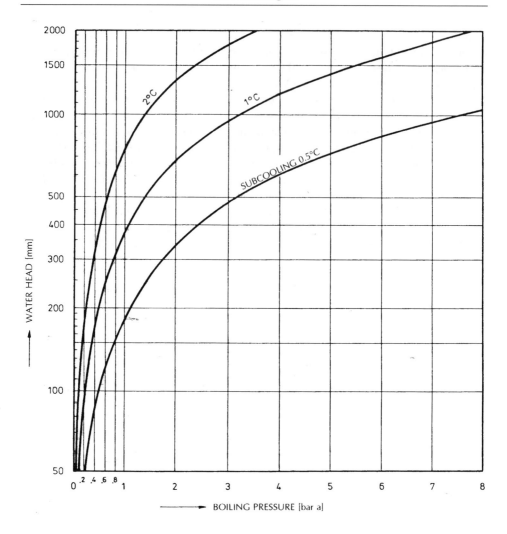

Fig. 1.18 Subcooling water head equivalent.

1.3 VESSEL SHELL

Apart from the stresses already mentioned, the pressure shell is subjected to additional strain by each weld joint. Wall weld joints are strain restrains to a greater or lesser extent.

Support rings acting as a reinforcement under vacuum conditions are no longer welded to the shell; they are installed unattached and held in position by brackets. Boiler supports are not connected to the shell of larger vessels; the tank or vessel is placed unattached in the boiler support saddles and held in the required position by brackets.

In upright heater shells the brackets are constructed with the shortest possible projection to reduce the moments acting on the shell to a minimum. The buckling resistance of the shell decreases as the diameter of the vessel increases. A calculation can be carried out in order to establish the breadth and height of the connection section required to absorb the moments from the supports As the thermal expansion is only restrained to a small degree by the supports, circumferential welds should be avoided as

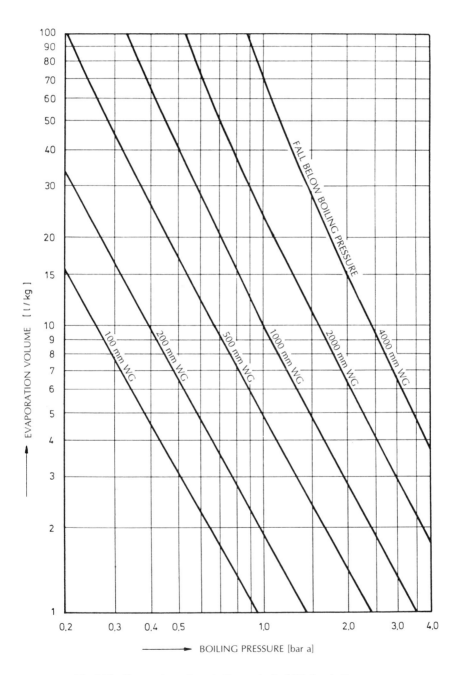

Fig. 1.19 Evaporation volume in the event of a fall below boiling pressure.

far as possible. The fastenings welded on in a longitudinal direction are either joined to the shell with continuous fillet welds or, in the case of heavy components, with double bevel butt joints (Fig. 1.20).

The aforementioned applies in particular to the feedwater tank because of its significance with respect to safety. Furthermore, it is particularly important that pressure vessels in nuclear power stations can be tested without difficulty. Vacuum rings are not used here. The rings would render examinations more difficult in the sections of the plant in which they are located. Vacuum strength is provided solely by the wall thickness. It is not permissible to use thin shells with vacuum reinforcement to absorb the pressure instead of thick shells because increased stresses would again occur in the transition section.

The feedwater tank steaming system requires several welds on the pressure shell. Here, too, the rule applies: only to the extent absolutely necessary, over the shortest possible distance and in an axial direction if possible. Each weld can be the cause of crevice corrosion induced by thermal expansion. Particular attention should therefore be paid to the arrangement and the form of the weld.

1.3.1 Feedwater Tank and Deaerator

The original function of the tank to store feedwater, has been pushed into the background as development has continued. The contrary is true for the low guarantee value of 0.005 mg/litre oxygen stipulated at the tank outlet: this requires a constant steam supply to

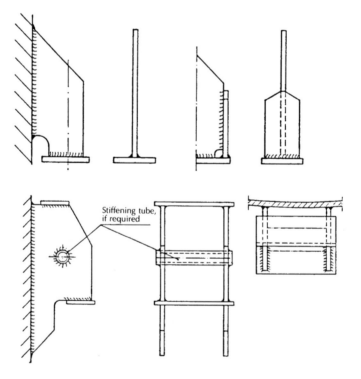

Fig. 1.20 Construction of the brackets.

maintain the boiling state in the entire feedwater tank. Using the trickle deaerator it became difficult to cope with the continuously increasing main condensate flow. The obvious solution was therefore to move the deaeration process into the feedwater tank already being charged with steam. The trickle element was changed to spray-type internals, which could be suspended in the storage tank, to cope with the greater flow rate. As deaeration took place under variable pressure, which means minimal steam pressure drops, the feedwater tank also became a combined heater with a good terminal temperature difference which is used as a storage tank for the feed pump.

A precondition for good deaeration is, first of all, the fine distribution of water for combined heating above the water level. Whilst passing through the steam space the temperature of the droplets approaches boiling temperature. The main quantity of oxygen is discharged here because the solvent power for the gas decreases as the temperature rises. This oxygen and the gas emerging from the surface of the water is at first dissolved spontaneously on the inflowing cold water droplets until saturation point is reached. This means that the entire spray cone is saturated with oxygen according to its temperature. The gas discharged from the heated droplets continuously dissolves again in the cold droplets and stabilises the degree of saturation. Only the quantity of gas exceeding saturation can be extracted. In the section containing the cold droplets where the gas is transported by the steam pressure, the partial pressure rises and the oxygen which can no longer be dissolved must be extracted. The inevitably required entrained steam flow and the losses become lower the closer the outlet is located to the cold face of the spray cone.

Residual deaeration, also referred to as boiling out, takes place in the lower section of the tank. Deaeration has been completed when the water flowing to the outlet nozzle is in a boiling state. This state cannot be attained; it is, however, possible by appropriately arranging the steam supply system and subsequent steam distribution to approach the ideal state so that the required 0.005 mg/litre O_2 content is then achieved. Only the water covering the steam bubble is in the boiling state and it is only here that the gas is released into the steam and transported to the spray cone of inflowing water. The residual O_2 content rises as the distance from the steam bubble and thus from the boiling point increases. Completely and partially deaerated water therefore flows to the outlet nozzle as a mixture. The required residual content in the mixture must comply with the set values. Cold make-up water saturated with oxygen should not be sprayed directly into the storage tank because the internals are not always able to intercept the subcooled strands which form to a greater extent in this way before they reach the outlet nozzle. If such a strand reaches the outlet, then the remaining, well deaerated water flow is not sufficient to reach an adequately low residual oxygen content in the mixture. The make-up water should always be pre-deaerated in a condenser or a separate trickle deaerator.

The steam supply system consists of one or several main tubes which are located above the regulated water surface and which extend over almost the entire length of the tank. A large number of tap lines, the lower ends of which are perforated to allow the steam to pass through, lead off the main tubes. The perforated ends are made of stainless steel to prevent the boreholes from being eroded. Wear occurs above all when wet steam flows through the boreholes at a high rate. When there is a high level of steam moisture, e.g., in nuclear power plants, the entire system should be made of stainless steel because the main tubes with their many branches are then also at risk. The weld joint between the C-steel and the stainless steel must always be under the water level. Weld joints between different materials located above the water would be alternately heated by superheated steam and subjected to shocks by boiling water. The different expansion coefficients of C-steel and stainless steel would soon lead to alternating stress damage.

If the heating steam is superheated, which applies at least when the storage tank pressure is sustained with steam from the reheating section, the expansion differences between the steam supply system and the vessel shell are to be taken into account. Tank lengths of 40 m and more are not unusual and temperature differences of 200°C between the main tube and the tank definitely possible. The expansion difference at the ends is therefore approx. 50 mm. Whilst the main tube is supported freely, the tap lines are fixed at the ends in the direction of the tank axis. The outer tap lines must therefore have the necessary flexibility over their length in order to compensate for expansion differences. The following measures are appropriate here:

1. select the diameter of the tap lines as small as possible
2. provide several steam inlets, to shorten the expansion paths
3. pre-stress the outer tap lines and thus halve the bending in the tubes. (Fig. 1.2.1).

Although the low moment of inertia and the great length of the tap lines allow for thermal expansion, the susceptibility to vibrations on the other hand increases. The system becomes unsusceptible to excitation when several tubes are interconnected without its capacity to compensate for expansion decreasing to any great extent.

The quantity of steam entering the tank is determined by the quantity and the temperature of the inflowing water. The jet cross-section of the main steam supply system is designed so that, even at the lowest steam rate, distribution over the length of the tank is assured, i.e., that the selected cross-section should be the largest possible for economic reasons and the smallest required for functional reasons. As a result, the pressure drop is low but this also applies to the outlet flow rate. During service the tank contents are always at boiling temperature and therefore the flow rate at the outlet is of no significance here. If the water were subcooled at the opening, because the tank contents have not yet reached boiling temperature, then considerable condensation shocks would constantly occur until this temperature has been reached.

The perforation in the service steam system is designed to allow large steam quantities, not available for heating, to flow through. This means that the tank contents cannot be heated by the service steam system when starting up the plant. A separate heating system is required, consisting of a main tube located above the water level with individual tap lines leading downwards. The design of this system is based on the pressure and the temperature of the steam available for initial heating and the required heating period. The pressure level should be such that one-and-a-half to twice the critical pressure difference is available. This pressure difference is necessary so that the steam emerging from the jet at a critical velocity maintains a steam space in the water ahead of the jet. The size of the steam space is determined by the steam supply and the condensation rate. Since the steam space is compressible it absorbs the condensation shocks; greater steam supply volumes and more superheated steam increase the capability of the steam space to do this. The critical velocity becomes excessively high downstream of the jet due to expansion and therefore the dispersion rate is initially higher than the condensation rate, which is reduced by the superheated state of the steam. Shortly thereafter, the steam or thermal rate and the condensation rate are the same and condensation shocks occur as a result of the disintegration of the steam jet. These condensation shocks are cushioned by the steam volume upstream of the jet. The higher the pressure and temperature of the initial heating steam, the less noise and vibration is caused by the initial heating process.

In order to avoid condensation shocks, the rule applies to all steam supply systems: horizontal tubes may only be located above the water level and the tap lines must be inclined to such an extent that the initial steam can displace the water without dispersing the cold water pocket.

The feedwater tank is of considerable significance as far as safety is concerned due to the energy stored therein. The operating loads also place the material under considerable stress. Statistics relating to damage show certain areas to be critical and it is essential to subject such areas to a surface crack examination during inspections. The critical areas are indicated in Fig. 1.21.

1.4 MULTISTAGE HEATER

It is always difficult to route the piping of the two lower steam extraction points (e.g., 4 × diameter 800 mm and 1 × diameter 1200 mm) of high capacity turbines to two different LP heater locations. This was the reason for developing so called Duplex heater which are two heaters combine in one shell. Two heating stages are located immediately adjacent to the condenser or even better in the condenser nozzle. Two steam shells are placed inside one another in this construction. The same applies to the water boxes. With the straight-tube construction, only the water flow is distributed to two tubesheets and two water boxes, the separation of the stages being designed similarly to that of a four-pass vessel.

Space restrictions were a determining factor for Duplex heaters but the design has now been modified reducing the number of tube fasteners and water boxes.

The principle of the multistage heater is as follows: the heat exchanger tube penetrates several steam spaces before ending in a water box. For the simplest type of multistage vessel, the two-stage or Duplex heater, the number of tube fasteners has been halved and only one tubesheet and one water box are required. This simplification is clearly illustrated in Fig. 1.22. On the left is the conventional heater train comprising individual vessels and on the right the corresponding multistage unit. The steam spaces are separated off with a partition plate through which the tubes pass from one stage to the next. The tubes are expanded into the boreholes in the 30 mm thick partition plate to ensure that the spaces are sealed off from one another. It was not possible to use this type of construction until a process had been developed whereby tubes can be accurately expanded at a predetermined point at a great tube depth. Using the hydraulic process (HYTEX), a tube can now pass through several heating sections which remain completely sealed off from one another.

Fig. 1.22 illustrates, among other things, how 2 U-tube heaters are transformed into a multistage U-tube heater. This transformation is taken one step further in Fig. 1.23 in a Duplex heater. The partition wall is circular and separates the outer space of steam extraction system A1 from the central space of extraction system A2. The circular arrangement of the U-tubes results in even distribution of the stress in the tubesheet and provides impact protection having wear-resistant stainless steel tubes at the edge. These tubes are located outside for the steam from A1 and they protect the centre from the steam from A2.

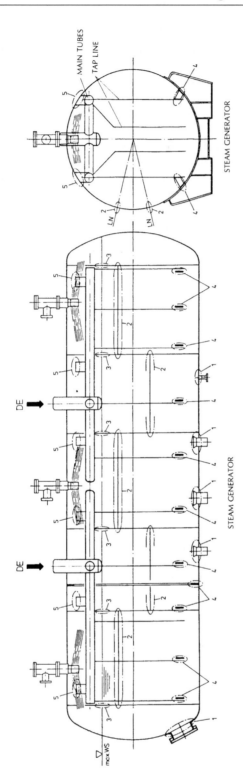

Fig. 1.21 Feedwater tank with deaeration elements and test field markings.

Critical area 1 = Nozzle welds below the water level
Critical area 2 = Longitudinal welds including butt joints in the sprayer section
Critical area 3 = Circumferential welds in the water/steam transition section
Critical area 4 = Attachments welded on to the vessel wall below the water level (e.g., vacuum rings)
Critical area 5 = Attachments welded on to the vessel wall above the water level (steam section)

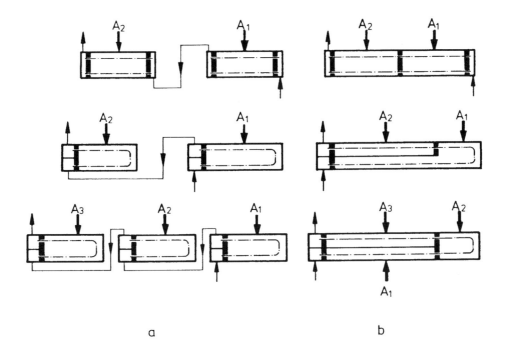

Fig. 1.22 Multistage heater: "a", conventional series connection; "b", multistage vessels.

1.5 THE HEAT EXCHANGER FOR PROCESS HEAT RECOVERY

The most important stage in heat recovery is the evaporation or steam generation process. Here the heat discharged in the process and transformed into steam is returned to the process through condensation; it may also be used to drive the auxiliary equipment required in the system, or to generate electricity.

If steam is merely transformed, i.e., if condensing higher pressure steam generates low pressure steam or, as in the case of the pressurised water reactor, the water on the primary side is cooled by steam generation on the secondary side, then the primary and secondary temperatures are close to each other.

The temperatures are considerably further apart, if another primary side medium is involved, in particular, a gas. The measures required to cope with these temperature differences in the process heat exchanger therefore take on completely different dimensions.

1.5.1 U-tube Steam Generator

The tubesheet with the large number of tube fastenings constitutes a critical component here, as it does in all heat exchangers. Fig. 1.24 is a diagram of a gas-heated steam generator operated in the once-through forced flow system on the water side. In constructions of this type the gas is cooled, for example from 500°C and 230 bar to approx. 330°C, by evaporating water. The water evaporates at approx. 310°C and 100 bar. The temperature difference of 190°C on the gas inlet side indicates that this section is particularly at risk.

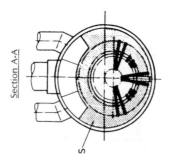

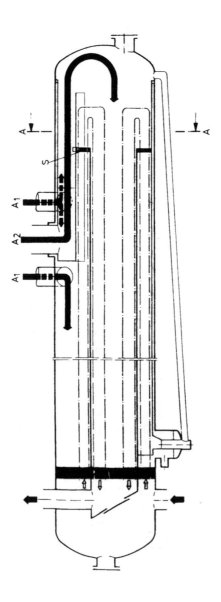

Fig. 1.23 Duplex heater of a composite construction: "S" partition wall.

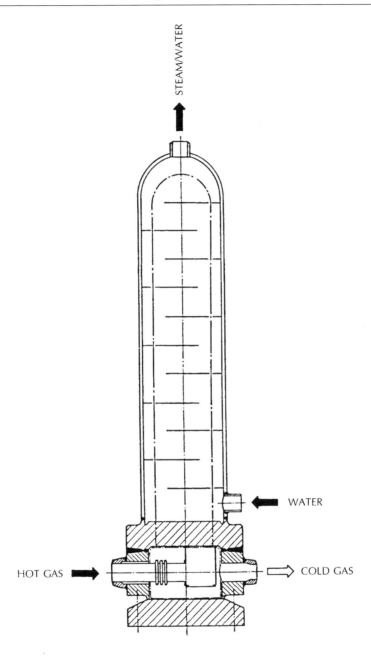

Fig. 1.24 Gas-heated steam generator (diagram).

What is more, many process gases place a significant strain on the material in the higher temperature and pressure ranges. This applies in particular to hydrogen and nitrogen.

The hot gas is conveyed so that it does not come into contact with the water box shell which the cold gas maintains at its own lower temperature. When the gas contains hydrogen, the tubesheets and water box material are selected according to the Nelson diagram [7] and when the gases contain nitrogen it is clad with austenitic material to prevent an increase in the nitrogen content. The material should be of step-cooling quality and the hardness of the sections in contact with the gases should be below 225 HB (Hardness Brinell) after manufacture.

The tubes are welded in and hydraulically expanded without a gap. Gap-free expansion up to the rear side of the tubesheet is essential to avoid gaps where concentration can occur. Hydraulic deformation leaves the least inherent stresses which is an important factor when hydrogen is involved [7].

On the gas inlet side, the tubes are provided with ferrules to prevent shock-related stresses and to reduce the temperature level. Despite these measures it must be borne in mind that cooling in the tubesheet is not sufficient to prevent nitrogen from building up at the tube ends and also on the borehole walls. The tubes which pass through the tubesheet keep the tubesheet at a virtually uniform temperature level despite the ferrules in the hot U-tube bend section. Over a distance of just a few millimetres, there is a sharp temperature drop to evaporation temperature just before the gas reaches the vicinity of the evaporating water. Widely differing methods have been used to achieve greater safety here.

The bundle shown in Fig. 1.25 consists of tubes arranged in a circle to ensure that the tubesheet has a good, rotationally symmetrical heat load. The tubes are welded to the water side of the tubesheet; good heat transfer to the outside of the tubes reduces the temperatures of the tubes to a level at which an increase in the nitrogen concentration is not possible (350 to 375°C). This is helped by the insulation installed between the outside of the ferrules and the inside of the U-tubes. The type of tube fastening selected here ensures that there is no possibility of the tube being affected by an increase in nitrogen concentration. This is essential for this tube concept because it is extremely difficult to repair a tube/tubesheet joint. The cold gas stream considerably reduces the increase in nitrogen concentration at the tubesheet/borehole joint but cannot prevent it completely. That is, however, a lesser problem and further developments have also been made here. Cooling boreholes were made in the tubesheet on the waterside (Fig. 1.26.). A centrally placed downcomer brings about circulation which discharges the heat from the tubesheet and, when the evaporation temperature is low enough, maintains the tube field borehole temperature below the nitration point. Consequently the level of hydrogen saturation is also lowered and, above all, the damage caused by H_2 oversaturation as a result of a temperature drop is reduced.

An interesting solution is alternate field tubing (Fig. 1.27). The hot and cold legs of the U-tubes alternate in the tube arrangement. That means that a hot tube end is always surrounded by cold tube ends and vice versa. A ferrule with appropriate intermediate insulation is also provided here. Fig. 1.28 shows the construction of a hot and cold tube end.

Due to the high pressures, thick tubesheets are required for U-tube bundles. This involves problems which are adequately resolved by the measures described above. Naturally, the problems posed by thick tubesheets do not arise with a thin tubesheet construction. The more robust thin tubesheet is integrated in a special tubesheet construction with appropriate internals giving it the stability of the thick tubesheet construction. Fig. 1.29 illustrates a section of the tubesheet with the support elements. High ligaments absorb the pressure load transmitted by the bolts as individual forces. The bolts as well as the tubes are welded to the tubesheet without any clearance.

All of the constructions previously mentioned are susceptible to the formation of deposits (wastage and denting) unless the water has been treated so that it is absolutely free from solids. Denting can be avoided by providing the tubes with a grid-type support. The header construction is one way of avoiding the consequences of deposits. Fig. 1.30 shows a header construction consisting of just one header. In the header, a guide element directs the hot gas through the ferrules directly into the bundle tubes (Fig. 1.31). The cold gas flows around the guide element thus insulating the header from the high temperature of the element.

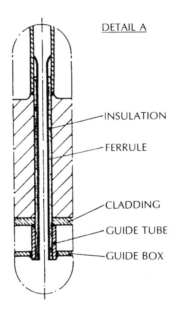

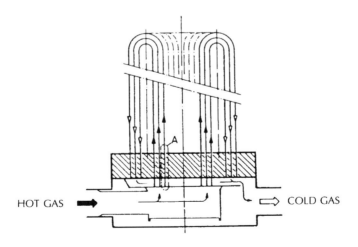

Fig. 1.25 Steam generator bundle with U-tubes arranged in a circle (diagram).

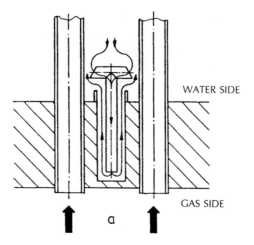

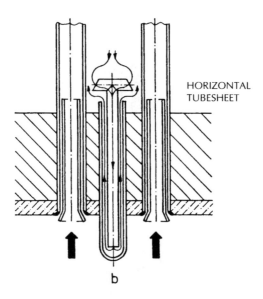

Fig. 1.26 Cooling boreholes: "a", pocket type; "b", pocket tube type.

1.5.2 Straight-tube Steam Generator

As an example, a horizontal type steam generator is selected which generates steam at approximately 310°C and 100 bar, but in this case with gas at 850 to 900°C and 30 bar. The vessel is installed horizontally and its flash tank is supported by downcomers and risers. The steam is generated in natural circulation (Fig. 1.32).

Two tubesheets form the separation between the hot gas and the evaporating water. The tubes through which the gas flows are pressure-sealed to these tubesheets. The central bypass tube regulates the temperature so that the gas at the outlet reaches the required temperature of approximately 500°C.

Although the evaporation section can be designed for the prevailing pressure and temperature in a conventional manner, the pressure walls of the inlet and outlet headers may not come into contact with the high temperature gas. Otherwise it would be necessary to use high-quality austenite similar to that used for the high temperature components for nuclear plants. Material costs and complicated manufacturing processes would not be economically justifiable and in this case not necessary as the medium gas allows the pressure shells and the tubesheets to be provided with internal insulation as protection against the high temperature.

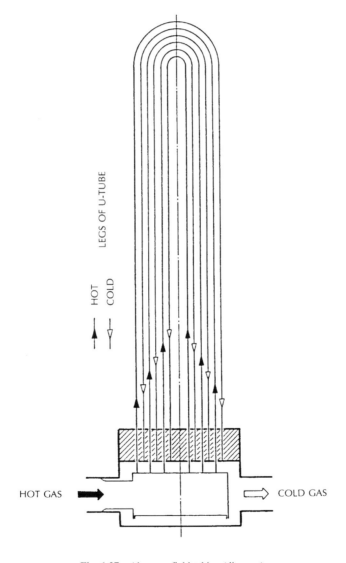

Fig. 1.27 Alternate field tubing (diagram).

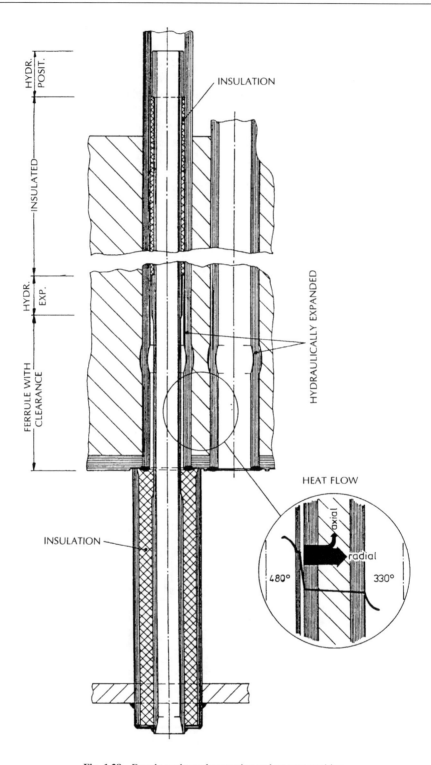

Fig. 1.28 Ferrule as thermal protection and as pass partition.

The thickness of the insulation which in this case consists of deoxidation-resistant aluminium oxide is selected so that the water box shell temperature remains below 200°C. The integrity of the insulation is monitored by regulating this temperature.

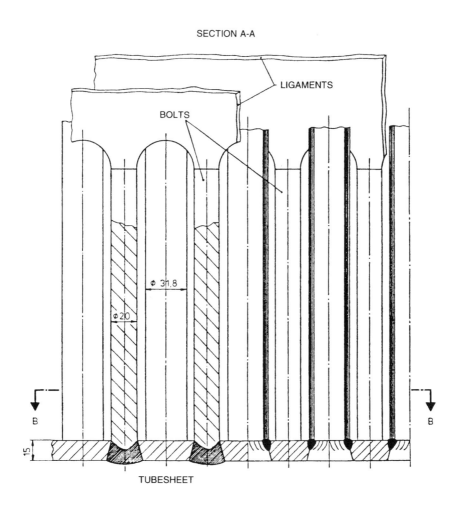

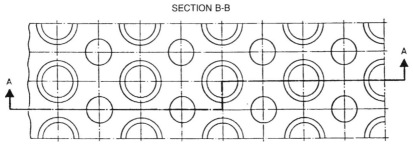

Fig. 1.29 Thin plate construction with tubes and bolts welded in without any clearance.

According to Nelson [7] it is possible to use nonalloyed steel if the temperature is reduced to 200°C and the partial pressure of the hydrogen is approximately 20 bar. The extent to which this advantage is used is determined by safety policy. Generally, the design temperature is increased to, e.g., 450°C, in order to safeguard against damage to the insulation. By insulating the tubesheet and using ferrules, the tubesheet temperature is maintained at 350°C so that nitration cannot occur.

The transition between the gas box shell, the tubesheet, and the evaporator shell is another important factor justifying the temperature reduction. This section cannot be subjected to a temperature difference of 500°C because the material strain here is already very high.

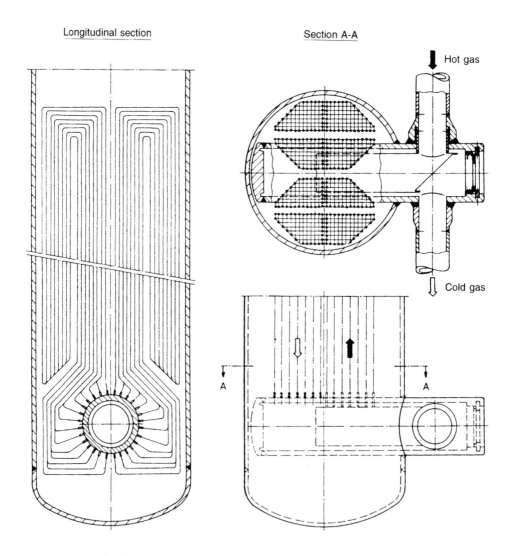

Fig. 1.30 Header construction of a steam generator with only one header.

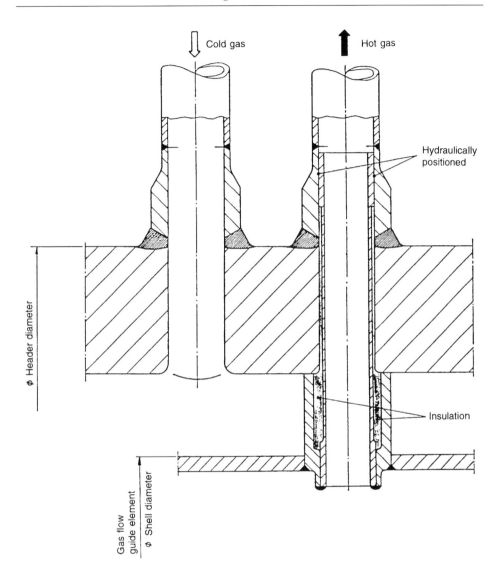

Fig. 1.31 Hot and cold gas passes in the header.

Various methods are used to reduce the strain on the critical tubesheet section but the thin tubesheet is still regarded as essential.

No exceptional stress is to be anticipated at the point at which the temperatures of the water box, the tubesheet, and the shell converge as expansion is virtually unrestricted and the transition radii are large. The expansion behaviour of the tubes is, however, to be assessed differently to the axial elongation of the shell. The shell does not have an expansion joint because of the high pressure. This means that the tubesheet which is kept thin for other reasons, inevitably takes on the role of a compensation element. The tubesheet is subject to double deformation constraint, first of all absorbing the elongation resulting from the temperature difference and then the elongation of the tubes as they absorb the load and act as an anchor between the tubesheets. The tubes therefore elongate more than the evaporator shell causing increased bending stress in the tubesheet contact section. The greatest surface stress occurs in the section between the tubesheet and the evaporator section shell (Fig. 1.33).

Basically, the stability of the annular zone is guaranteed as shown in Fig. 1.33. The protective layer is the reason for using other, more complicated constructions to relieve the load under some circumstances.

Even if the material-related admissible stresses are not exceeded at a point on a wetted ferritic component subject to high stress, such as this annular section, the strains involved are frequently too great for the protective layer. This discrepancy often arises when taking advantage of the higher strength of alloyed steels. Load cycles such as those which occur during start-up and shutdown can cause the magnetite layer to tear, exposing the material to a risk of cracking.

The protective layer forms only during the operating phase. A stress-free oxide layer rests on a surface which is generally under tensile stress. The protective layer can therefore be damaged only when the vessel is shut down, because the contracting surface places the adherent magnetite layer under compressive stress, thus upsetting the outer layer. Therefore, only the maximum elongation at a higher pressure load can normally be considered when assessing the service life.

It has been proposed on many occasions that a magnetite layer should be built up in the vessel before it is put into operation. This, however, makes sense only if the vessel is under pressure.

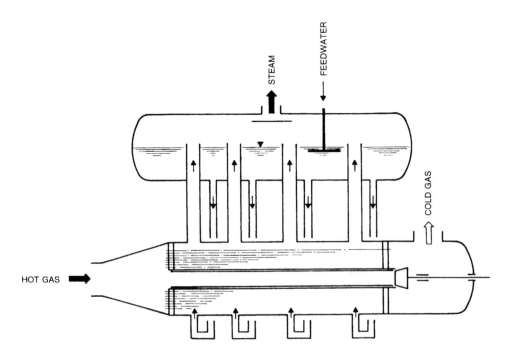

Fig. 1.32 Straight-tube steam generator with steam drum.

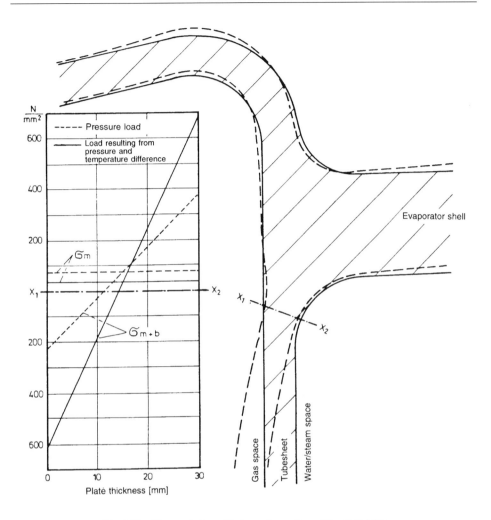

Fig. 1.33 Stresses in the critical annular section of the plate joint.

Then it could be done again during the operating phase. A protective layer which builds up on a base material which is not under pressure is worse than no protective layer at all.

Magnetite and hydrogen are produced during the formation of the protective layer according to the formula:

$$3Fe + 4H_2O = Fe_3O_4 + 4H_2$$

The greater the stress of the steel surface, the more hydrogen it absorbs from the protective layer formation process. The surface also becomes more susceptible to cracking. The Schikorr reaction [1] occurring after damage to the surface enables an incipient crack to become larger, even without load changes which accelerate cracking.

The buildup of a protective layer which is necessary and inevitable under correct operation means that the design engineer cannot apply high quality steels in the normal way. The elongation of such steels cannot be fully utilised This fact not only restricts the effective application of the material but must also be taken into account in the construction.

The problem of tubesheet restraint has been resolved in two different ways: by adapting the construction and by adapting the method of manufacture.

Fig. 1.34 illustrates the constructional solution. The details are already familiar from Fig. 1.29. The thin tubesheet is supported by elements which do not restrict its thermal function. The support structure prevents negative deformation of the tubesheet and eliminates undesirable stress in the throat section.

The manufacturing process permits a completely different solution. In this case the deformation to which the thin tubesheet is subject during operation is reduced to a minimum by prestressing the tubes during manufacture.

With a conventional non-prestressed construction, the tubes cannot start to absorb the load until the elongation caused by the difference in temperature between the shell and the tubes is completed.

This does not apply to peripheral tubes. These are not subject to tensile stress because they are adjacent to the shell, but they exert pressure on the circumference of the tubesheet which even intensifies the deformation.

Using the Strain Blocking System (SBS) [8] the thermal elongation, which can be accurately determined can be eliminated by appropriate prestressing. In this way, the deformation pressure of the tubesheet is retarded from the start when the system is put into operation. Prestressing can, of course, exceed the thermal elongation balance or, in cases where there is no temperature difference, be used merely to reduce deformation stresses as a result of tensile or pressure load.

For a steam generator of the type discussed here, the finite element (FE) calculation shows the prestressing to be adequate to compensate for the temperature difference. This is clearly illustrated in the computer drawings in Fig. 1.35. The effects of the operating loads are shown individually in sections 1, 2 and 4. Section 3 illustrates the predeformation obtained by prestressing the tubes. Section 5 is obtained by combining the other 4 sections and it illustrates well the operating situation and the good deformation balance achieved by using the SBS. (The $\Delta l = 2.3$ mm in section 3 relates to a tube half length of 3 m).

When using SBS it is necessary to verify the effect of the induced prestressing for the test pressure if it was based on a high thermal elongation difference. At ambient temperature, the test pressure could inadmissible increase the prestressing and render modifications necessary.

Fig. 1.36 illustrates how the SBS is applied. The tubes have already been hydraulically expanded into tubesheet 1. Tube a can still move freely in tubesheet 2; tube b was already secured before reaching the specified Δl after being heated with a heating cartridge.

The following situation arises after blocking:

- All the tube ends in tubesheet 2 project by exactly Δl. The required elongation is therefore blocked in the tubes and reserved for the temperature difference under operating conditions.
- The tube ends are in the welding position in tubesheet 2 or, as in the example, in the sleeves.
- The vessel must not be heat treated after the blocking process. All heat treatment required due to the material or the wall thickness must have been already carried out.

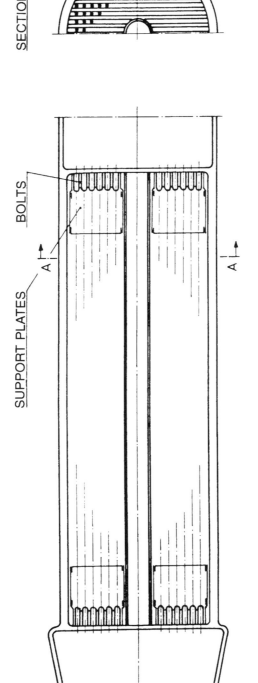

Fig. 1.34 Straight-tube steam generator with a tubesheet support structure.

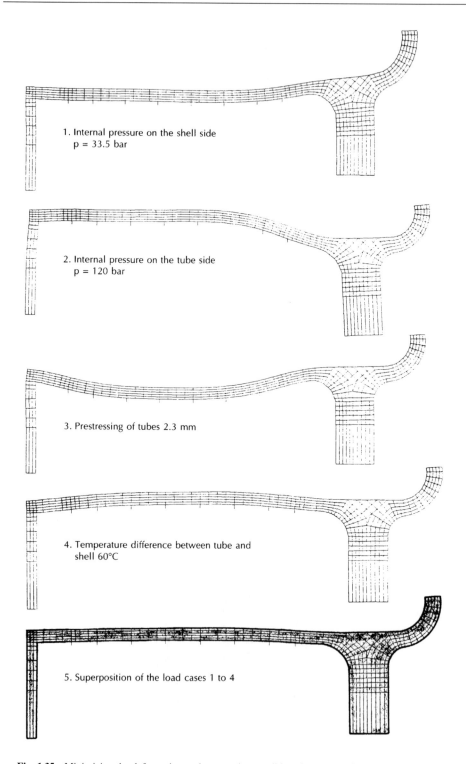

Fig. 1.35 Minimizing the deformation under operating conditions by prestressing the tubes with SBS.

Heat Exchangers for Power Stations

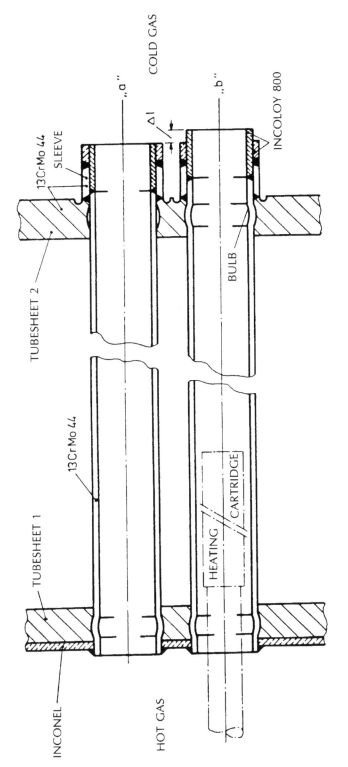

Fig. 1.36 Prestressing of the tubes: "a", not fastened; "b", prestressed and fastened.

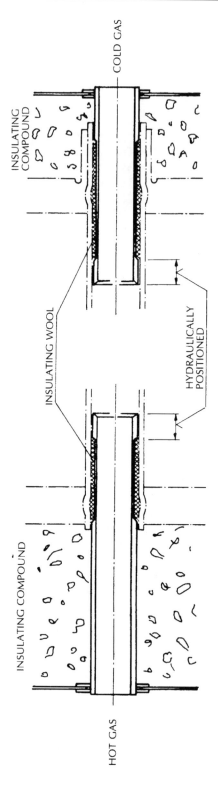

Fig. 1.37 Thermal protection of the tubesheets.

Sleeves are provided if the tubes are made of material requiring heat treatment. The sleeves and the tubes are tipped with a material which can be welded without requiring heat treatment.

The tube ends are not welded without a gap but hydraulically positioned without any clearance. The advantage of this is that the tube/tubesheet weld is not subjected to tube forces, all the welds can be fully examined and repairs are easy to carry out.

Fig. 1.37 illustrates the thermal protection of the tube sheets. Here it is important to position the ends of the tube inserts against the inside wall of the tube so that the gas cannot flow behind them.

Finally, it should be noted that it is very difficult to carry out low stress annealing on a straight-tube heat exchanger. The shell leads with its temperature both in the heating and the cooling processes. The stresses should die down during the holding period, provided that the tubes do not lag behind the shell by more than 50 to 100°C during heating. Temperature lags as low as 50 to 100°C can generate undesirable residual stresses during cooling. If it is necessary to limit these temperature spreads, this can be done by extending the annealing time, which is uneconomic, or by charging the tubes with a side-stream from the furnace. The latter process involves considerable control procedures.

The problem does not even arise if SBS is applied in manufacturing steam generators. The tubes retain the freedom to expand during the entire heat treatment period and are completely relieved of stress before being blocked.

BIBLIOGRAPHY

1. Effertz et al., Korrosion und Erosion in Speisewasser-Vorwärmern - Ursache u. Verhütung, *Der Maschinenschaden*, 51 (1978), Heft 4.
2. Achten, N. et al., Die Bedeutung von Spalten zwischen Rohren und Rohrboden für das betriebliche Verhalten von Wärmetauschern, *Chem. Ing.-Techn.*, 58 (1986), Nr. 9.
3. Green, S. and Paine, J. P. N., Materials performance in nuclear pressurized water reactor steam generators, *Nuclear Technology*, Vol. 55, Okt. 81.
4. Pettigrew, M. J., Sylvester, Y., and Campagna, A. O., Vibration Analysis of Heat Exchangers and Steam Generator Designs, *Nucl. Eng. and Design*, 48 (1978).
5. Pettigrew, M. J., Tromp, J. H., and Mastorakos, J., Vibration of Tube Bundles Subjected to Two-Phase Cross-Flow, *Journal of Pressure Vessel Technology*, 1985, Vol. 107.
6. Podhorsky, M., Die rechnerischen Schwerpunkte bei der Konstruktion und die festigkeitsmäßige Auslegung einer Vorwärmestrecke, *VGB-KRAFTWERKSTECHNIK*, 64 (1984), Heft 1.
7. Stellwag, B. and Kaesche, H., Kinetik der wasserstoffinduzierten Spannungsrißkorrosion, *Werkstoffe und Korrosion*, 33 (1982).
8. Podhorsky, M., Design of Modern Heat Exchangers Using Hydraulic Tube Expansion, 6-IC PVT, Beijing, September 1988.
9. TEMA, Standards of the Tubular Exchanger Manufacturers Association, TEMA, New York.
10. Spence, J. and Tooth, A. S., Eds., Pressure Vessel Design and Principles, E&FN Spon, Chapman & Hall, London, 1994.
11. Singh, K. P. and Soler, A. I., Mechanical Design of Heat Exchangers and Pressure Vessels Components, Arcturus Publishers, New Jersey, 1984.
12. HEDH, The Heat Exchanger Design Handbook, Begell House, New York.
13. Pressure Vessel and Piping Design, Collected Papers, 1927–1959, ASME, New York, 1960.

2
CALCULATION OF STRUCTURAL STRESSES USING THE FORCE METHOD

2.1 INTRODUCTION

Heat transfer constructions are made up of simple geometric structural elements. Not only can such individual elements be manufactured economically but they can also withstand loads extremely effectively.

Internal pressure is the most significant primary load which determines the geometry of a vessel. For this reason it is important that the pressure-bearing enclosure is shaped according to one of the simplest shells of revolution, i.e., a conical or cylindrical shell or a circular (toroidal) shell, which is often referred to as the knuckle.

In most constructions the two pressure spaces are separated by a flat tubesheet or a thick-walled cylindrical header.

The structural elements mentioned here are normally dimensioned according to recognised technical codes. It is a known fact that the majority of the German and other technical codes take account only of internal pressure load in the continuous sections; they neglect the discontinuity stresses which occur between the individual structural elements due to the different shapes which are not always matched. The magnitude of these stresses can sometimes be such that alternating plastic deformation or progressive deformation can occur. The notch impact stresses at the transitions play a significant role as far as material fatigue is concerned. With high strength steels there is a particularly high risk that discontinuity stresses will exceed the admissible limits. It is therefore advisable to calculate these stresses and to provide the appropriate safety margins.

The force method is one way of determining discontinuity stresses. With this method, the entire vessel structure is broken down into individual structural components, for which the main differential equations and the corresponding solutions can be given. One lateral force and one moment are introduced in the sections as statically non-determined variables. These unknown section variables are determined using the compatibility condition which requires the uniformity of displacement and inclination in the assumed section. The task can be reduced to the solution of a linear equation system, in which the section variables represent unknown values and the corresponding vectors of the given loads are placed on the righthand side. Two further equations are then necessary for each assumed section.

The force method is ideal for quickly determining discontinuity stresses in rotationally symmetrical constructions. It is relatively easy to calculate rotationally symmetrical loads and the discontinuity stress can therefore be checked quickly and easily during the construction phase using a desktop computer or a pocket calculator.

Peak stresses in notches and transitions cannot be calculated using the force method. It is, however, possible to form a general picture about the fatigue in a construction of a detail by using the stress-increase factors indicated in the relevant literature and obtained from strain gauge measurements or previously carried out calculations.

A considerably greater amount of calculation work is required for non rotationally symmetrical loads and discontinuity stresses, because then a Fourier method has to be used and the number of unknown quantities increases by the number of the Fourier coefficients.

2.2 APPLICATION OF THE FORCE METHOD IN THE CONSTRUCTION OF HEAT TRANSFER EQUIPMENT

The structural elements of heat transfer equipment are ideal for the application of the force method. Typical applications are the calculation of flanges [1] and the calculation of the pressure shell of a U-tube or a straight tube heat exchanger [2].

The basic elements required for the majority of heat transfer equipment constructions are:

- flat heads
- dished heads
- spherical shell
- cylindrical shell
- conical shell
- tubesheet
- ring

The deformation equations for the individual basic elements are set out below without derivation so that it is subsequently possible to draw up the linear equation system for the various combinations without any difficulty.

Elasticity matrices are specified for the boundary loads, for the internal pressure, and for the constant temperature.

The following relationship can be established between the deformations and the loads:

$$\{v\} = [N]\{f\}$$

v = deformation vector
N = elasticity matrix
f = force vector

The deformation equations are specified according to this system.

2.3 FLAT HEADS

This component is used only in special cases and only for small diameters. The reason is the non uniform material utilisation brought about by internal pressure load and the resultant high inherent weight. The flat head which is also referred to as a flat cover is a simple circular plate with a constant plate thickness. The deformation equations relating to loads caused by internal pressure, boundary loads and by a constant temperature change compared to the initial state are as follows:

Calculation of Structural Stresses Using the Force Method

$$\begin{bmatrix} w \\ \theta \end{bmatrix} = \begin{bmatrix} \dfrac{2F_3}{3E(t/R)} & \dfrac{F_3}{ER(t/R)^2} \\ \dfrac{F_3}{ER(t/R)^2} & \dfrac{2F_3}{ER^2(t/R)^3} \end{bmatrix} \begin{bmatrix} Q \\ M \end{bmatrix} + p_i \begin{bmatrix} -\dfrac{t}{2} \dfrac{F_1}{E(t/R)^3} \\ -\dfrac{F_1}{E(t/R)^3} \end{bmatrix} + \begin{bmatrix} \alpha(T_m - T_0)R_m \\ 0 \end{bmatrix} \quad (2.1)$$

The significance of the individual letters can be seen from Fig. 2.1 or is described in detail. E is the modulus of elasticity, α the temperature coefficient of expansion, T_0 is the initial temperature and T_m is the mean temperature of the structure at a given time.

The remaining unknown quantities have the following form:

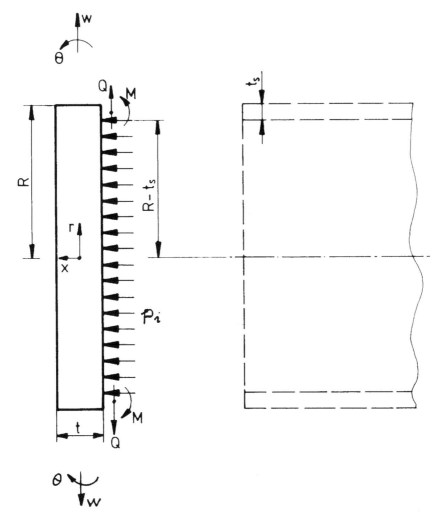

Fig. 2.1 Flat head.

$$R_m = R - \frac{t_s}{2}$$

$$f = \frac{t_s}{R}$$

$$F_1 = \frac{3(1-\mu)(2-f^2)(1-f^2)\{8-f(4-f)(1-\mu)\}}{16(2-f)}$$

$$F_2 = \frac{3}{8}(1-f^2)\left\{(1-\mu)(2-f^2) + 4(1+\mu)\left(1+2\ln\frac{2-f}{2-2f}\right)\right\}$$

$$F_3 = \frac{3}{8}(1-\mu)(2-f)\{8-f(4-f)(1-\mu)\}$$

$$F_4 = \frac{1}{8}\{8-f(4-f)(1-\mu)\} \tag{2.2}$$

The stresses are calculated using the following equations:

$$\sigma_r = \frac{F_4}{t}\left[1-\frac{6x}{t}\right]Q - \frac{12 \cdot F_4 \cdot x}{t^3}M + \frac{xp_i}{t(t/R)^2}\left[F_2 - \frac{3(3+\mu)r^2}{4R^2}\right]$$

$$\sigma_t = \frac{F_4}{t}\left[1-\frac{6x}{t}\right]Q - \frac{12 \cdot F_4 \cdot x}{t^3}M + \frac{xp_i}{t(t/R)^2}\left[F_2 - \frac{3(1+3\mu)r^2}{4R^2}\right]$$

$$\sigma_a = \left(x-\frac{t}{2}\right)\frac{p_i}{t} \tag{2.3}$$

2.4 DISHED HEADS

The dished head comprises a spherical shell and a circular shell, also referred to as the knuckle. This combination produces a very light-weight head construction which is particularly suitable for large diameters.

Two types of dished heads have become established:

- torispherical head: $R = Da$; $r = 0.1\ Da$; $0.001 \leq \frac{s}{D_a} \leq 0.1$
- elliptical head: $R = 0.8\ Da$; $r = 0.154\ Da$; $0.001 \leq \frac{s}{D_a} \leq 0.1$

The equations for the deformation calculation given here have been kept general so that they apply for the two most frequent types. The dished head is broken down into two elementary shells, the spherical shell and the circular shell. Corresponding differential equations are relatively simple to resolve for the spherical shell and for the loads acting on it.

The calculation of the deformation of the circular shell is carried out using the force method according to [3]. For this, the knuckle is broken down into a finite number of elements. The breakdown into 4 elements illustrated in Fig. 2.2 is adequately accurate in practice and has the advantage that the variables of the appropriate computer program are effectively reduced.

The deformation equations of the spherical shell subjected to internal pressure p_i, boundary loads Q_0, M_0 according to Fig. 2.2 and temperature change compared to the initial state T_0 can be set up as follows:

$$\begin{bmatrix} w_0 \\ \theta_0 \end{bmatrix} = \begin{bmatrix} \dfrac{\cos\varphi_0 \, W_a}{E\varphi_0} & \dfrac{W_b}{Et\varphi_0} \\ \dfrac{\cos\varphi_0 \, X_a}{Et\varphi_0} & \dfrac{X_b}{Et^2\varphi_0} \end{bmatrix} \begin{bmatrix} Q_0 \\ M_0 \end{bmatrix}$$

$$+ p_i \begin{bmatrix} -\left(\dfrac{R\cos^2\varphi_0}{2\varphi_0} W_a - \dfrac{1-\mu}{2} \dfrac{R^2}{t} \sin\varphi_0\right)\dfrac{1}{E} \\ -\dfrac{R\cos^2\varphi_0}{2\varphi_0 \, tE} X_a \end{bmatrix} + \begin{bmatrix} \alpha(T_m - T_0)R\sin\varphi_0 \\ 0 \end{bmatrix} \quad (2.4)$$

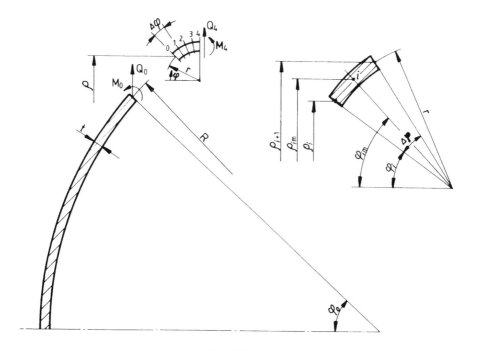

Fig. 2.2 Dished head.

with the individual abbreviations

$$W_a = \frac{4(K\varphi_0)^3 - 2(K\varphi_0)^2 + 3/8(K\varphi_0) - 9/8}{2(K\varphi_0)^2 - 1/2(K\varphi_0) - 3/4} \cdot \frac{(K\varphi_0)^2}{\sqrt{3(1-\mu^2)}}$$

$$W_b = \frac{4(K\varphi_0)^4}{2(K\varphi_0)^2 - 1/2(K\varphi_0) - 3/4}$$

$$X_a = \frac{4(K\varphi_0)^4 - 3(K\varphi_0) + 9/16}{2(K\varphi_0)^2 - 1/2(K\varphi_0)^2 - 3/4}$$

$$X_b = \frac{2(k\varphi_0)^3}{2(K\varphi_0)^2 - 1/2(K\varphi_0)^2 - 3/4} \cdot 4 \cdot \sqrt{3(1-\mu^2)}$$

$$K = \left[3(1-\mu^2)\right]^{1/4} \sqrt{\frac{R}{t}} \tag{2.5}$$

The tangential and meridian stresses at any point on the spherical shell can be expressed as follows:

$$\sigma_N = \frac{N}{t} \mp \frac{6M}{t^2}$$

$$\sigma_r = \frac{T}{t} \mp 6\mu \frac{M}{t^2} \tag{2.6}$$

whereby the + sign applies to the inside of the shell and the – sign to the outside.

The individual internal forces and moments can be calculated from the boundary loads and from the internal pressure.

$$N = A*n_a + B*n_b + \frac{p_i(R-t/2)}{2}$$

$$T = A*t_a + B*t_b + \frac{p_i(R-t/2)}{2}$$

$$\frac{M}{t} = A*m_a + B*m_b \tag{2.7}$$

The abbreviations signify the following expressions:

$$A^* = A_1 \cos\varphi_0 Q_0 + A_2 \frac{M_0}{t} - A_1 \cos^2\varphi_0 \frac{p_i(R-t/2)}{2}$$

$$B^* = B_1 \cos\varphi_0 Q_0 + B_2 \frac{M_0}{t} - B_1 \cos^2\varphi_0 \frac{p_i(R-t/2)}{2} \tag{2.8}$$

The unknown quantities not yet explicitly specified n_a, n_b, t_a, t_b, m_a, m_b, A_1, A_2, B_1, B_2 are indicated in Tables 2.1 and 2.2. These tables were taken from [3].

The knuckle is not suitable for simple deformation equations as these involve idealisation with elements. The deformation equations for the i-element are as follows:

TABLE 2.1 Coefficients for the Integration Constants of the Spherical Shell Subject to Stress on the Outer Edge [3]

$K\varphi_0$	$K\sqrt{2}\,\varphi_0$	A_1	A_2	B_1	B_2
0.07	0.10	+ 0.004	+ 3.794	+ 1.414	− 0.004
0.14	0.20	+ 0.013	+ 3.794	+ 1.413	− 0.019
0.28	0.40	+ 0.073	+ 3.793	+ 1.413	− 0.075
0.57	0.80	+ 0.292	+ 3.765	+ 1.393	− 0.301
0.85	1.20	+ 0.641	+ 3.654	+ 1.313	− 0.663
1.13	1.60	+ 1.066	+ 3.365	+ 1.112	− 1.108
1.41	2.00	+ 1.461	+ 2.852	+ 0.313	− 1.533
1.70	2.40	+ 1.695	+ 2.151	+ 0.155	− 1.815
1.98	2.80	+ 1.694	+ 1.397	− 0.231	− 1.877
2.26	3.20	+ 1.478	+ 0.698	− 0.657	− 1.735
2.55	3.60	+ 1.129	+ 0.171	− 0.939	− 1.460
2.83	4.00	+ 0.735	− 0.177	− 1.065	− 1.133
3.11	4.40	+ 0.383	− 0.373	− 1.059	− 0.811
3.39	4.80	+ 0.053	− 0.451	− 0.953	− 0.527
3.68	5.20	− 0.167	− 0.450	− 0.787	− 0.297
3.96	5.60	− 0.323	− 0.399	− 0.594	− 0.126
4.24	6.00	− 0.394	− 0.324	− 0.401	− 0.008
4.40	6.22	− 0.404	− 0.281	− 0.303	+ 0.034
4.60	6.51	− 0.393	− 0.224	− 0.192	+ 0.069
4.80	6.79	− 0.367	− 0.171	− 0.098	+ 0.090
5.00	7.07	− 0.330	− 0.125	− 0.022	+ 0.101
5.20	7.35	− 0.283	− 0.085	+ 0.037	+ 0.103
5.40	7.64	− 0.234	− 0.052	+ 0.080	+ 0.098
5.60	7.92	− 0.184	− 0.026	+ 0.107	+ 0.088
5.80	8.20	− 0.137	− 0.007	+ 0.122	+ 0.076
6.00	8.49	− 0.095	+ 0.007	+ 0.126	+ 0.063
6.50	9.19	− 0.016	+ 0.023	+ 0.106	+ 0.033
7.00	9.90	+ 0.025	+ 0.022	+ 0.068	+ 0.011
7.50	10.61	+ 0.036	+ 0.015	+ 0.032	− 0.000
8.00	11.31	+ 0.031	+ 0.008	+ 0.007	− 0.005
8.50	12.02	+ 0.020	+ 0.003	− 0.005	− 0.005
9.00	12.73	+ 0.010	+ 0.000	− 0.010	− 0.003
9.50	13.44	+ 0.002	+ 0.001	− 0.009	− 0.002
10.00	14.14	+ 0.001	+ 0.001	− 0.006	− 0.000

TABLE 2.2 Coefficients for the Pattern of the Internal Forces in the Spherical Shell Subjected to Stress on the Outer Edge [3]

Kφ	K√2 φ	n_a		n_b		t_a		t_b		m_a		m_b
0.07	0.10	+	0.0008	+	0.7071	+	0.0027	+	0.7071	+	0.2635	− 0.0009
0.14	0.20	+	0.0035	+	0.7071	+	0.0106	+	0.7071	+	0.2635	− 0.0034
0.28	0.40	+	0.0141	+	0.7070	+	0.0424	+	0.7066	+	0.2634	− 0.0137
0.57	0.80	+	0.0566	+	0.7056	+	0.1696	+	0.6990	+	0.2612	− 0.0548
0.85	1.20	+	0.1271	+	0.6995	+	0.3802	+	0.6690	+	0.2516	− 0.1228
0.13	1.60	+	0.2250	+	0.6830	+	0.6698	+	0.5868	+	0.2258	− 0.2165
1.41	2.00	+	0.3487	+	0.6484	+	1.0274	+	0.4147	+	0.1762	− 0.3329
1.70	2.40	+	0.4917	+	0.5859	+	1.4254	+	0.1065	+	0.0752	− 0.4618
1.98	2.80	+	0.6563	+	0.4844	+	1.8228	−	0.3722	−	0.0809	− 0.5923
2.26	3.20	+	0.8235	+	0.3314	+	2.1485	−	1.1296	−	0.3121	− 0.7019
2.55	3.60	+	0.9813	+	0.1153	+	2.2994	−	2.1456	−	0.6310	− 0.7587
2.83	4.00	+	1.1081	−	0.1736	+	2.1341	−	3.4515	−	1.0420	− 0.7189
3.11	4.40	+	1.1759	−	0.5410	+	1.4727	−	5.0144	−	1.5352	− 0.5265
3.39	4.80	+	1.1191	−	0.9846	+	0.1007	−	6.7273	−	2.0789	− 0.1155
3.68	5.20	+	0.9865	−	1.4915	−	2.2108	−	8.3804	−	2.6097	+ 0.5856
3.96	5.60	+	0.6116	−	2.0314	−	5.6759	−	9.6304	−	3.0228	+ 1.6943
4.24	6.00	+	0.0691	−	2.5565	−	10.4419	−	9.9705	−	3.1632	+ 3.1080
4.40	6.22	−	0.350	−	2.841	−	13.83	−	9.58	−	3.080	+ 4.14
4.60	6.51	−	1.005	−	3.160	−	18.73	−	8.29	−	2.703	+ 5.65
4.80	6.79	−	1.821	−	3.381	−	24.10	−	5.74	−	1.925	+ 7.33
5.00	7.07	−	2.809	−	3.437	−	29.89	−	1.43	−	0.593	+ 9.10
5.20	7.35	−	3.945	−	3.261	−	35.56	+	5.21	+	1.427	+ 10.85
5.40	7.64	−	5.221	−	2.788	−	40.64	+	14.53	+	4.246	+ 12.43
5.60	7.92	−	6.559	−	1.961	−	44.43	+	26.73	+	7.941	+ 13.63
5.80	8.20	−	7.900	−	0.717	−	46.04	+	42.02	+	12.582	+ 14.17
6.00	8.49	−	9.145	+	1.006	−	44.31	+	60.46	+	18.187	+ 13.71
6.50	9.19	−	10.040	+	7.708	−	16.14	+	118.01	+	35.751	+ 5.36
7.00	9.90	−	8.841	+	17.788	+	67.08	+	177.51	+	54.064	− 19.67
7.50	10.61	+	1.143	+	29.508	+	229.31	+	197.99	+	60.777	− 68.75
8.00	11.31	+	22.674	+	37.936	+	473.55	+	103.12	+	32.82	− 142.94
8.50	12.02	+	57.333	+	33.753	+	745.58	−	217.32	−	63.27	− 226.13
9.00	12.73	+	100.626	+	3.240	+	884.40	−	878.19	−	262.56	− 269.95
9.50	13.44	+	136.610	−	69.036	+	573.7	−	1919.1	−	577.71	− 178.98
10.00	14.14	+	132.487	−	192.469	−	666.0	−	3153.3	−	953.47	+ 191.86

$$\frac{w_{i+1}}{r} = H_2 \frac{w_i}{r} + H_3 Q_i + H_4 \frac{M_i}{Etr} + H_5 \frac{M_{i+1}}{Etr}$$

$$+ H_6 \frac{Q_{i+1} + Q_i}{Et} + H_7 \frac{p_i}{E} + \alpha(T_m - T_0) \frac{p_{im}}{r}$$

$$\theta_{i+1} = \theta_i + H_1 \frac{M_{i+1} + M_i}{Etr} \tag{2.9}$$

The section variables Q and M can be expressed as follows for the i-element:

Calculation of Structural Stresses Using the Force Method

$$\frac{Q_{i+1}}{Et} = (F_1 + F_2)\frac{Q_i}{Et} + F_3\frac{w_i}{r} + F_4\theta_i - (F_5 + F_6)\frac{p_i}{E}$$

$$\frac{M_{i+1}}{Etr} = G_0\frac{M_i}{Etr} - (G_1 + G_2)\frac{Q_i}{Et} - G_3\frac{w_i}{r} - G_4\theta_i$$

$$+ (G_5 + G_6 + G_7)\frac{p_i}{E} \qquad (2.10)$$

In turn, the stresses in the knuckle can be simply calculated as follows:

$$\sigma_N = \frac{N}{t} \mp \frac{6M}{t^2}$$

$$\sigma_T = \frac{T}{t} \mp \frac{6\mu M}{t^2}$$

with

$$N = Q\cos\varphi + \frac{p_i \rho}{2}\sin\varphi$$

$$T = \frac{Esw}{\rho} + rN \qquad (2.11)$$

The abbreviations used in the equations are as follows:

$$F_1 = \frac{p_i}{p_{i+1}}; \qquad F_4 = F_3(\cos\varphi_i - \cos\varphi_{im})$$

$$F_2 = \frac{r}{4p_{i+1}}\Delta\varphi\cos\varphi_i; \qquad F_5 = \frac{(r-t/2)p_{im}}{tp_{i+1}}\Delta\varphi\sin\varphi_{im}$$

$$F_3 = \frac{r^2}{p_{im}p_i}\Delta\varphi; \qquad F_6 = \frac{rp_i}{8tp_{i+1}}\Delta\varphi\sin\varphi_i$$

$$G_0 = F_1; \qquad G_4 = F_3\frac{\Delta\varphi^2}{6}\sin^2\varphi_{im}$$

$$G_1 = F_1\Delta\varphi\sin\varphi_{im}; \qquad G_5 = \frac{p_i^2}{p_{i+1}t}\frac{\Delta\varphi}{2}\cos\varphi_{im}$$

$$G_2 = F_2(\cos\varphi_{im} - \cos\varphi_{i+1}); \quad G_6 = \frac{(r-t/2)p_{im}}{tp_{i+1}}\frac{\Delta\varphi^2}{2}$$

$$G_3 = F_3(\cos\varphi_{im} - \cos\varphi_{i+1}); \quad G_7 = F_6(\cos\varphi_{im} - \cos\varphi_{i+1})$$

$$H_1 = \frac{r^2}{t^2}6\Delta\varphi; \qquad H_5 = \frac{r^2}{t^2}2\Delta\varphi^2\left(\sin\varphi_{im} + \frac{\Delta\varphi}{4}\cos\varphi_{im}\right)$$

$$H_2 = 1 - \frac{r}{4\rho_i}\Delta\varphi\cos\varphi_{im}; \qquad H_6 = \frac{1-\mu^2}{2}\Delta\varphi\cos^2\varphi_{im}$$

$$H_3 = \Delta\varphi\sin\varphi_{im}; \qquad H_7 = \frac{1-\mu^2}{2}\frac{p_{im}}{t}\Delta\varphi\sin\varphi_{im}\cos\varphi_{im}$$

$$H_4 = \frac{r^2}{t^2}2\Delta\varphi^2\left(2\sin\varphi_{im} + \frac{\Delta\varphi}{4}\cos\varphi_{im}\right) \tag{2.12}$$

Now it is possible to calculate the cross-section variables between the individual elements and thus to determine the deformation of the knuckle.

2.5 SPHERICAL SHELLS

The appropriate equations for the spherical shell were dealt with in the previous section. For a hemispherical head which is often used in practice and which has very favourable stress patterns and good material utilization, $\phi_0 = 90°$ would have to be used in the equations (2.4 to 2.8). These equations are, however, quite complex for this case and too complicated for the limited purpose of calculating a hemispherical head.

For this reason simple equations are indicated below. These are well suited for programming.

The deformation equations are then:

$$\begin{bmatrix} w \\ \theta \end{bmatrix} = \begin{bmatrix} \frac{2R\lambda}{Et} & \frac{2\lambda^2}{Et} \\ \frac{2\lambda^2}{Et} & \frac{4\lambda^2}{ERt} \end{bmatrix} \begin{bmatrix} Q \\ M \end{bmatrix} + \begin{bmatrix} \frac{2R^3(1-2\mu)+(R+t/2)^3(1+\mu)}{2ER^2(u^3-1)} \\ 0 \end{bmatrix} P_i$$

$$+ \alpha(T_m - T_0)R\begin{bmatrix} 1 \\ 0 \end{bmatrix} \tag{2.13}$$

with the abbreviations

$$\lambda = \beta \cdot R$$

$$\beta = \left[3(1-\mu^2)\right]^{1/4}\frac{1}{\sqrt{Rt}}$$

Calculation of Structural Stresses Using the Force Method

$$u = \frac{R+t/2}{R-t/2} \qquad (2.14)$$

The following are also required to calculate the section variables and the deformations along the hemispherical head:

$$D = \frac{Et^3}{12(1-\mu^2)}$$

$$x = \frac{\pi R}{180}(90-\varphi)$$

$$f_1(\beta x) = e^{-\beta x} \cos\beta x$$
$$f_2(\beta x) = e^{-\beta x} (\cos\beta x - \sin\beta x)$$
$$f_3(\beta x) = e^{-\beta x} (\cos\beta x + \sin\beta x)$$
$$f_4(\beta x) = e^{-\beta x} \sin\beta x \qquad (2.15)$$

The equations indicated below describe the convergence of the section variables along the spherical head:

$$Q(x) = Q \sin\varphi \, f_2(\beta x) - 2\beta M f_4(\beta x)$$

$$M(x) = Q \frac{\sin\varphi}{\beta} f_4(\beta x) + M f_3(\beta x)$$

$$w(x) = Q \frac{\sin^2\varphi}{2\beta^2 D} f_1(\beta x) + M \frac{\sin\varphi}{2\beta^2 D} f_2(\beta x) \qquad (2.16)$$

and the stresses can be expressed as:

$$\sigma_N = \pm \frac{6M(x)}{t^2} + \sigma_{pi}$$

$$\sigma_T = \frac{Ew(x)}{R} \pm \mu \frac{6M(x)}{t^2} + \sigma_{pi}$$

with

$$\sigma_{pi} = \frac{p_i(R-t/2)}{2t} \qquad (2.17)$$

The + sign applies to the inside of the hemispherical shell and the − sign to the outside. The equations given here are valid up to a ratio of R/t > 10.

With a smaller ratio the results are inaccurate in practice and should be treated with caution. This also applies to the other shells.

2.6 CYLINDRICAL SHELLS

Fig. 2.3 shows the designations introduced here for the geometrical variables. The cylindrical shell always has two edges which can either interact with one another, in which case it is referred to as a short cylindrical shell, or the two edges are so far apart that no interaction can take place. In this case they are long cylindrical shells with correspondingly simple deformation equations.

The general deformation equations applicable for both short and long cylindrical shells are indicated.

$$\begin{bmatrix} w_L \\ \theta_L \\ w_R \\ \theta_R \end{bmatrix} = \begin{bmatrix} -\dfrac{B_{11}}{2\beta^3 D} & \dfrac{B_{12}}{2\beta^2 D} & -\dfrac{G_{11}}{2\beta^3 D} & \dfrac{G_{12}}{2\beta^2 D} \\ -\dfrac{B_{12}}{2\beta^2 D} & \dfrac{B_{22}}{2\beta D} & -\dfrac{G_{12}}{2\beta^2 D} & \dfrac{G_{22}}{2\beta D} \\ -\dfrac{G_{11}}{2\beta^3 D} & \dfrac{G_{12}}{2\beta^2 D} & -\dfrac{B_{11}}{2\beta^2 D} & \dfrac{B_{12}}{2\beta^2 D} \\ -\dfrac{G_{12}}{2\beta^2 D} & \dfrac{G_{22}}{2\beta D} & -\dfrac{B_{12}}{2\beta^2 D} & \dfrac{B_{22}}{2\beta D} \end{bmatrix} \begin{bmatrix} Q_L \\ M_L \\ Q_R \\ M_R \end{bmatrix} +$$

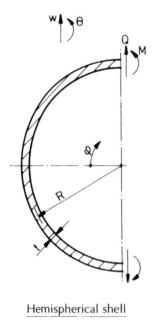
Hemispherical shell

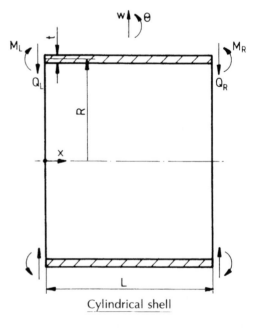
Cylindrical shell

Fig. 2.3 Cylindrical shell.

$$+ \ p_i \begin{bmatrix} \dfrac{(1-\mu/2)R(R-t/2)}{Et} \\ 0 \\ \dfrac{(1-\mu/2)R(R-t/2)}{Et} \\ 0 \end{bmatrix} + \alpha(T_m - T_0)R \begin{bmatrix} 1 \\ 0 \\ 1 \\ 0 \end{bmatrix}$$

(2.18)

The abbreviations for the short cylindrical shell are:

$$B_{11} = \frac{\sinh 2\beta L - \sin 2\beta L}{2(\sinh^2 \beta L - \sin^2 \beta L)}$$

$$B_{12} = \frac{\cosh 2\beta L - \cos \beta L}{2(\sinh^2 \beta L - \sin^2 \beta L)}$$

$$B_{22} = \frac{\sinh 2\beta L + \sin 2\beta L}{2(\sinh^2 \beta L - \sin^2 \beta L)}$$

$$G_{11} = - \frac{\cosh \beta L \sin \beta L - \sinh \beta L \cos \beta L}{\sinh^2 \beta L - \sin^2 \beta L}$$

$$G_{12} = - \frac{\sinh \beta L \sin \beta L}{\sinh^2 \beta L - \sin^2 \beta L}$$

$$G_{22} = - \frac{\cosh \beta L \sin \beta L + \sinh \beta L \cos \beta L}{\sinh^2 \beta L - \sin^2 \beta L} \quad (2.19)$$

and for the long cylindrical shell

$$B_{11} = B_{12} = B_{22} = 1$$
$$G_{11} = G_{12} = G_{22} = 0 \quad (2.20)$$

The displacement at a distance x from the lefthand edge of the section is calculated as follows:

$$w_{(x)} = - \frac{F_{11}(\beta x)}{2\beta^3 D} Q_L + \frac{F_{12}(\beta x)}{2\beta^2 D} M_L + \frac{F_{13}(\beta x)}{\beta} \theta_L + F_{14}(\beta x) w_L \quad (2.21)$$

The variables θ_L and w_L refer to inclination or displacement in the section indicated as a result of, for example, different temperatures but not as a result of section variables Q_L and M_L.

$$F_{11}(\beta x) = (\cosh\beta x \sin\beta x - \sinh\beta x \cos\beta x)/2$$

$$F_{12}(\beta x) = \sinh\beta x \sin\beta x$$

$$F_{13}(\beta x) = (\cosh\beta x \sin\beta x + \sinh\beta x \cos\beta x)/2$$

$$F_{14}(\beta x) = \cosh\beta x \cos\beta x \quad (2.22)$$

The moment in the axial direction at a distance x from the lefthand edge of the section can be calculated as follows

$$M(x) = 2\beta^2 D \left[-\frac{F_{13}(\beta x)}{2\beta^3 D} Q_L + \frac{F_{14}(\beta x)}{2\beta^2 D} M_L - \frac{F_{11}(\beta x)}{\beta} \theta_L - F_{12}(\beta x) w_L \right] \quad (2.23)$$

whereby D represents the cylinder rigidity and β the convergence factor of the cylindrical shell

$$D = \frac{Et^3}{12(1-\mu^2)}$$

$$\beta = \left[3(1-\mu^2)\right]^{1/4} \frac{1}{\sqrt{Rt}} \quad (2.24)$$

The axial and tangential stresses can be expressed as follows:

$$\sigma_a = \pm \frac{6M(x)}{t^2} + \frac{p_i(R-t/2)}{2t}$$

$$\sigma_t = \frac{Ew(x)}{R} \pm 6\mu \frac{M(x)}{t^2} + \frac{p_i(R-t/2)}{t} \quad (2.25)$$

They can now be calculated in any section along the cylindrical shell. The + sign applies to the inside, the − sign to the outside.

2.7 CONICAL SHELLS

In the case of a pressure-bearing shell of a pressure vessel, the conical shell frequently forms the transition between two cylindrical shells with different diameters. Therefore it occurs almost exclusively as a double-edged conical shell. A single-edge conical shell can also be regarded as a special type of double-edged conical shell. The geometric variables are indicated in Fig. 2.4.

The deformation vector which describes the deformation of the two edges as a result of the unknown boundary variables Q and M relating to both internal pressure and a temperature change from the initial temperature T_0 to temperature T_m is:

$$\{w\} = [L_2][L_1]^{-1}\{F\} + p_i\{a\} + \alpha(T_m - T_0)\{b\} \tag{2.26}$$

No derivation is made here.

The meaning of the individual vectors and their abbreviations is as follows:

$$\{w\} = \begin{Bmatrix} w_R \\ \theta_R \\ w_L \\ \theta_L \end{Bmatrix} \tag{2.27}$$

$$\{F\} = \begin{Bmatrix} Q_R \\ M_R \\ Q_L \\ M_L \end{Bmatrix} \tag{2.28}$$

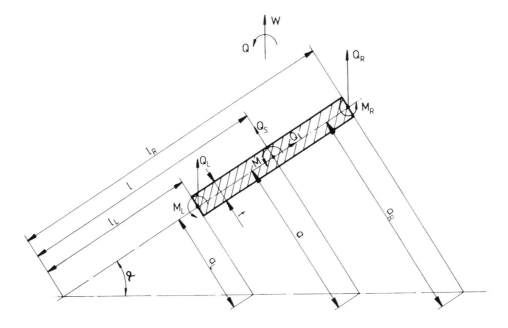

Fig. 2.4 Conical shell.

$$\{a\} = p_i \begin{Bmatrix} \dfrac{\cos\alpha}{2a_R^2 Et}\left[(1-2\mu)(a_R-t/2)^2 a_R^2 + (1+\mu)(a_R-t/2)^2(a_R+t/2)^2\right] \\ \dfrac{tg\alpha}{a_R Et}(a_R^2-t^2/4)\left[3/2+(1+\mu)\dfrac{t^2}{4a_R^2}\right] \\ \dfrac{\cos\alpha}{2a_L^2 Et}\left[(1-2\mu)(a_L-t/2)^2 a_L^2 + (1+\mu)(a_L-t/2)^2(a_L+t/2)^2\right] \\ \dfrac{tg\alpha}{a_L Et}(a_L^2-t^2/4)\left[3/2+(1+\mu)\dfrac{t^2}{4a_L^2}\right] \end{Bmatrix}$$

(2.29)

$$\{b\} = \begin{Bmatrix} a_R \cos\alpha \\ 0 \\ a_L \cos\alpha \\ 0 \end{Bmatrix}$$

(2.30)

The matrices can be set out as follows:

$$[L_1] = \begin{bmatrix} -\dfrac{A_1}{e_1} & -\dfrac{A_2}{e_1} & -\dfrac{A_3}{e_1} & -\dfrac{A_4}{e_1} \\ -\dfrac{D_1}{e_2} & -\dfrac{D_2}{e_2} & -\dfrac{D_3}{e_2} & -\dfrac{D_4}{e_2} \\ \dfrac{A_1}{e_3} & \dfrac{A_2}{e_3} & \dfrac{A_3}{e_3} & \dfrac{A_4}{e_3} \\ \dfrac{D_1}{e_4} & \dfrac{D_2}{e_4} & \dfrac{D_3}{e_4} & \dfrac{D_4}{e_4} \end{bmatrix}$$

$$[L_2] = -\begin{bmatrix} F_1 g_1 & F_2 g_1 & F_3 g_1 & F_4 g_1 \\ G_1 g_2 & G_2 g_2 & G_3 g_2 & G_4 g_2 \\ F_1 g_3 & F_2 g_3 & F_3 g_3 & F_4 g_3 \\ G_1 g_4 & G_2 g_4 & G_3 g_4 & G_4 g_4 \end{bmatrix}$$

(2.31)

The meanings of the individual coefficients are:

$$e_1 = I_R \cos\alpha \quad ; \quad g_1 = \dfrac{I_R \sin\alpha}{Et}$$

$$e_2 = \xi(I_R)/2 \quad ; \quad g_2 = \dfrac{12\, I_R}{Et^3}$$

Calculation of Structural Stresses Using the Force Method

$$e_3 = I_L \cos \alpha \quad ; \quad g_3 = \frac{I_L \sin \alpha}{Et}$$

$$e_4 = \xi(I_L)/2 \quad ; \quad g_4 = \frac{12\, I_L}{Et^3}$$

$$\chi = \sqrt[4]{\frac{(1-\mu^2)\cot^2\alpha}{t^2}} \quad ; \quad \xi(I) = 2\,\chi\sqrt{I} \qquad (2.32)$$

$$A_1(\xi) = T_1(\xi) + \frac{2}{\xi}\, T_2'(\xi)$$

$$A_2(\xi) = T_2(\xi) - \frac{2}{\xi}\, T_1'(\xi)$$

$$A_3(\xi) = T_3(\xi) + \frac{2}{\xi}\, T_4'(\xi)$$

$$A_4(\xi) = T_4(\xi) - \frac{2}{\xi}\, T_3'(\xi)$$

$$B_1(\xi) = T_1''(\xi) - \frac{2}{\xi}\, T_1'(\xi) - \frac{4}{\xi^2}\, T_2'(\xi)$$

$$B_2(\xi) = T_2''(\xi) - \frac{2}{\xi}\, T_2'(\xi) + \frac{4}{\xi^2}\, T_1'(\xi)$$

$$B_3(\xi) = T_3''(\xi) - \frac{2}{\xi}\, T_3'(\xi) - \frac{4}{\xi^2}\, T_4'(\xi)$$

$$B_4(\xi) = T_4''(\xi) - \frac{2}{\xi}\, T_4'(\xi) + \frac{4}{\xi^2}\, T_3'(\xi)$$

$$D_1(\xi) = T_2'(\xi) - \frac{2(1-\mu)}{\xi}\, T_2(\xi) + \frac{4(1-\mu)}{\xi^2}\, T_1'(\xi)$$

$$D_2(\xi) = -T_1'(\xi) + \frac{2(1-\mu)}{\xi}\, T_1(\xi) + \frac{4(1-\mu)}{\xi^2}\, T_2'(\xi)$$

$$D_3(\xi) = T_4'(\xi) - \frac{2(1-\mu)}{\xi}\, T_4(\xi) + \frac{4(1-\mu)}{\xi^2}\, T_3'(\xi)$$

$$D_4(\xi) = -T_3'(\xi) + \frac{2(1-\mu)}{\xi}\, T_3(\xi) + \frac{4(1-\mu)}{\xi^2}\, T_4'(\xi)$$

$$E_1(\xi) = \mu T_2'(\xi) + \frac{2(1-\mu)}{\xi} T_2(\xi) - \frac{4(1-\mu)}{\xi^2} T_1(\xi)$$

$$E_2(\xi) = -\mu T_1'(\xi) - \frac{2(1-\mu)}{\xi} T_1(\xi) - \frac{4(1-\mu)}{\xi^2} T_2(\xi)$$

$$E_3(\xi) = +\mu T_4'(\xi) + \frac{2(1-\mu)}{\xi} T_4(\xi) - \frac{4(1-\mu)}{\xi^2} T_3(\xi)$$

$$E_4(\xi) = -\mu T_3'(\xi) - \frac{2(1-\mu)}{\xi} T_3(\xi) - \frac{4(1-\mu)}{\xi^2} T_4(\xi)$$

$$F_i(\xi) = -\mathrm{tg}\alpha\left[\frac{\mu}{I} A_i(\xi) - \frac{2\chi^2}{\xi} B_i(\xi)\right]$$

$$G_i(\xi) = -\frac{2}{\xi}\left[\mu D_i(\xi) - E_i(\xi)\right] \qquad (2.33)$$

With $i = 1 \ldots 4$ and further with the functions

$$T_1(\xi) = \mathrm{ber}\,\xi \quad ; \quad T_1'(\xi) = \mathrm{ber}'\,\xi$$

$$T_2(\xi) = -\mathrm{bei}\,\xi \quad ; \quad T_2'(\xi) = -\mathrm{bei}'\,\xi$$

$$T_3(\xi) = -\frac{2}{\pi}\mathrm{kei}\,\xi \quad ; \quad T_3'(\xi) = -\frac{2}{\pi}\mathrm{kei}'\,\xi$$

$$T_4(\xi) = -\frac{2}{\pi}\mathrm{ker}\,\xi \quad ; \quad T_4'(\xi) = -\frac{2}{\pi}\mathrm{ker}'\,\xi \qquad (2.34)$$

ber, bei, ker and kei being the Kelvin functions and ber', bei', ker' and kei' the derivations of the Kelvin functions according to ξ.

It is then relatively simple to calculate the displacement and the torsion of the conical shell at a distance I from the tip of the cone if the boundary variables of the conical shell are known from the force method overall equation system. The same applies to the cross-section variables. Index u was selected for the circumferential direction.

The applicable equations are:

$$w(I) = -\frac{I\sin\alpha}{Et}\left[f_1 F_1(\xi) + f_2 F_2(\xi) + f_3 F_3(\xi) + f_4 F_4(\xi)\right] +$$

$$+ \frac{p_i \cos\alpha}{2a^2 Et}\left[(1-2\mu)(a-t/2)^2 a^2 + (1+\mu)(a-t/2)^2(a+t/2)^2\right]$$

$$+ \alpha(T_m - T_0)a \cos\alpha$$

$$\theta(I) = -\frac{12\,I}{Et^3}\left[f_1 G_1(\xi) + f_2 G_2(\xi) + f_3 G_3(\xi) + f_4 G_4(\xi)\right]$$

$$+ \frac{p_i\,tg\alpha}{aEt}\left(a^2 - \frac{t^2}{4}\right)\left[\frac{3}{2} + (1+\mu)\frac{t^2}{4a^2}\right]$$

$$Q_s(I) = -\frac{1}{I}\left[f_1 A_1(\xi) + f_2 A_2(\xi) + f_3 A_3(\xi) + f_4 A_4(\xi)\right]$$

$$Q_1(I) = Q_s(I)\,tg\alpha + \frac{(a-t/2)^2 p_i}{2a}$$

$$M(I) = -\frac{2}{\xi(I)}\left[f_1 D_1(\xi) + f_2 D_2(\xi) + f_3 D_3(\xi) + f_4 D_4(\xi)\right]$$

$$Q_u(I) = -\frac{2\chi^2 tg\alpha}{\xi(I)}\left[f_1 B_1(\xi) + f_2 B_2(\xi) + f_3 B_3(\xi) + f_4 B_4(\xi)\right] + p_i(a-t/2)^2\frac{1}{a}$$

$$M_u(I) = -\frac{2}{\xi(I)}\left[f_1 E_1(\xi) + f_2 E_2(\xi) + f_3 E_3(\xi) + f_4 E_4(\xi)\right] \tag{2.35}$$

The integration constants f_1 to f_4 must be determined from the boundary conditions.

$$\{f\} = [L_1]^{-1} \cdot \{F\} \tag{2.36}$$

As is the case with the cylindrical shell, it is necessary to determine the area influenced by an edge. The influence of one edge on another can be ignored if the regression equation

$$\sqrt[4]{12(1-\mu^2)}\cot^2\alpha\left[\sqrt{I_R/t} - \sqrt{I_L/t}\right] \geq 2 \tag{2.37}$$

is fulfilled. The given criterion is met more easily by steeper and thinner shells and then the double-edged conical shell can be regarded as two single-edged conical shells with the half cone angles α or $(\pi-\alpha)$.

The formulas for the calculation of the stresses in the conical shell are simple and do not need to be given in detail.

2.8 TUBESHEET

The tubesheet is an extremely important structural element of a heat exchanger as it separates the two pressure sections and therefore the media. Using the force method the tubesheet can either be divided up into narrow rings (normally applied to heat exchanger construction models with straight tubes and two tubesheets) or as described below, an algorithm is used. In this case the tubesheet is considered together with the rigid ring, the tubesheet tubing being axially symmetrical. This calculation model is particularly suitable for the U-tube construction of a heat exchanger with different pressures and temperatures on the two sides.

The tubed section of the tubesheet is treated as a homogenous and isotropic circular plate with corrected modulus of elasticity E^* and corrected Poisson's ratio μ^* but with a constant plate thickness h. These corrected values are given, for example, in ASME III, A-8000, Fig. A-8131-1 as a function of the ligament efficiency. The temperature of the perforated plate is that of the tube-side medium, i.e., equal to T_L. The deformation equations which can be used to calculate the statically unknown boundary forces and boundary moments based on the boundary conditions are indicated below. The designation of the individual calculation variables is given in Fig. 2.5.

$$\begin{Bmatrix} w_L \\ \theta_L \\ w_R \\ \theta_R \end{Bmatrix} = \begin{bmatrix} A_1 + \dfrac{h^2}{4B} r_{m1} & \dfrac{h r_{m1}}{2B} & -A_1 + \dfrac{h^2}{4B} r_{m2} & \dfrac{h r_{m2}}{2B} \\ -\dfrac{h}{2B} r_{m1} & -\dfrac{r_{m1}}{B} & -\dfrac{h}{2B} r_{m2} & -\dfrac{r_{m2}}{B} \\ A_1 - \dfrac{h^2}{4B} r_{m1} & -\dfrac{h r_{m1}}{2B} & -A_1 - \dfrac{h^2}{4B} r_{m2} & -\dfrac{h r_{m2}}{2B} \\ -\dfrac{h}{2B} r_{m1} & -\dfrac{r_{m1}}{B} & -\dfrac{h}{2B} r_{m2} & -\dfrac{r_{m2}}{B} \end{bmatrix} \begin{Bmatrix} Q_L \\ M_L \\ Q_R \\ M_R \end{Bmatrix}$$

$$+ \, p_L A_2 \begin{Bmatrix} -h/2 \\ 1 \\ h/2 \\ 1 \end{Bmatrix} + p_R A_3 \begin{Bmatrix} h/2 \\ -1 \\ -h/2 \\ -1 \end{Bmatrix} + \begin{Bmatrix} \delta_L - \theta^* h/2 \\ \theta^* \\ \delta_R + \delta_{PT} + \theta^* h/2 \\ \theta^* \end{Bmatrix} \quad (2.38)$$

The abbreviations in (2.38) have the following form. No derivation is given. It can be found, for example, in [2].

$$r^* = r_0 + \dfrac{d_0}{4}$$

$$\bar{r}_R = \dfrac{4 r_1 - s_L - s_R}{4}$$

Calculation of Structural Stresses Using the Force Method

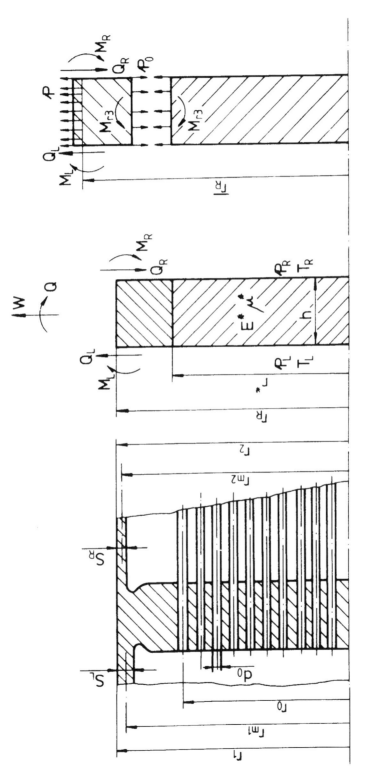

Fig. 2.5 Tubesheet.

Heat Exchangers

$$B = D_R + D_2(1+\mu^*)$$

$$D_R = \frac{Eh^3}{12}\ln\left(\frac{r_1}{r^*}\right)$$

$$D_2 = \frac{E^* h^3}{12(1-\mu^{*2})}$$

$$\frac{p}{p_o} = \frac{2\bar{r}_R^2}{r^{*2}(1-\mu)+\bar{r}_R^2(1+\mu)+(1-\mu^*)\left(\bar{r}_R^2 - r^{*2}\right)E/E^*}$$

$$p_o = \frac{(Q_L - Q_R)r_{m1}}{hr_R}$$

$$M_{r3} = (p_1 - p_2)\frac{r^{*2}}{8} - D_2(1-\mu^*)\frac{\theta_R}{r^*}$$

$$\theta^* = -\frac{r^* + r_R}{2}\frac{(T_L - T_R)}{h}\alpha + \frac{\delta_{PT}}{h}$$

$$\delta_L = \alpha r_{m1}\frac{T_L + T_R - 2T_0}{2} \quad;\quad \delta_R = \alpha r_{m2}\frac{T_L + T_R - 2T_0}{2}$$

$$\delta_{PT} = \frac{p_i}{E}\left[\frac{(1-\mu)r^{*2}\bar{r}_R + (1+\mu)r^{*2}r_R^2/\bar{r}_R}{r_R^2 - r^{*2}}\right]$$

$$\frac{p_i}{E} = \frac{\alpha(T_L - T_S)}{2}\frac{1}{\frac{r_R^2 + r^{*2}}{r_R^2 - r^{*2}} + \mu + \frac{E}{E^*}(1-\mu^*)}$$

$$A_1 = -\frac{r_{m1}}{Eh}\frac{2r^{*2}(p/p_o) - r_R^2(1-\mu) + r^{*2}(1+\mu)}{r_R^2 - r^{*2}}$$

$$A_2 = \frac{1}{2B}\left(\frac{r_1^3}{3} - \frac{r^{*3}}{12} - \frac{r_1^2 S_L}{2} - \frac{S_L^3}{6}\right)$$

$$A_3 = \frac{1}{2B}\left(\frac{r_2^3}{3} - \frac{r^{*3}}{12} - \frac{r_2^2 S_R}{2} - \frac{S_R^3}{6}\right) \tag{2.39}$$

It should also be mentioned that α is the designation for the thermal expansion coefficient. The stresses in the tubesheet can be calculated using the familiar formula for

Calculation of Structural Stresses Using the Force Method

a circular plate and a ring as indicated in the standard reference works. ASME, III A-8000 gives details about stress increases in the ligament and at the edge of the boreholes.

2.9 RINGS

The final basic element for the deformation equations given here is the ring. In the heat transfer equipment sector, the ring is used, for example, as a flange, a stiffener or for shell branches.

The deformation conditions on the classical flange ring are explained in Chapter 4.

The loads to which the ring is subjected comprise the unknown boundary forces and boundary moments (Fig. 2.6), the internal pressure p_i and the temperature change $(T_m - T_0)$. T_m represents the mean temperature at the time of the calculation and T_0 the mean initial temperature of the ring.

The geometric designations are given in Fig. 2.6. The deformation equations for which once again no derivation is made, are then:

$$\begin{Bmatrix} w_L \\ \theta_L \\ w_R \\ \theta_R \end{Bmatrix} = \begin{bmatrix} A+B & C & (-A+B) & -C \\ C & \dfrac{2C}{h} & -C & -\dfrac{2C}{h} \\ -(A-B) & -C & (A+B) & C \\ C & \dfrac{2C}{h} & -C & -\dfrac{2C}{h} \end{bmatrix} \begin{Bmatrix} Q_L \\ M_L \\ Q_R \\ M_R \end{Bmatrix}$$

$$+ \alpha(T_m - T_0) \begin{Bmatrix} r_L + \dfrac{s_L}{2} \\ 0 \\ r_R + \dfrac{s_R}{2} \\ 0 \end{Bmatrix}$$

$$+ p_i \begin{Bmatrix} \dfrac{h r_i^2 (a_L - a_R)}{4 E D_R} + \dfrac{r_i(r_a + r_i)}{2E(r_a - r_i)} \\ \dfrac{r_i^2 (a_L - a_R)}{2 E D_R} \\ -\dfrac{h r_i^2 (a_L - a_R)}{4 E D_R} + \dfrac{r_i(r_a + r_i)}{2E(r_a + r_i)} \\ \dfrac{r_i^2 (a_L - a_R)}{2 E D_R} \end{Bmatrix} \quad (2.40)$$

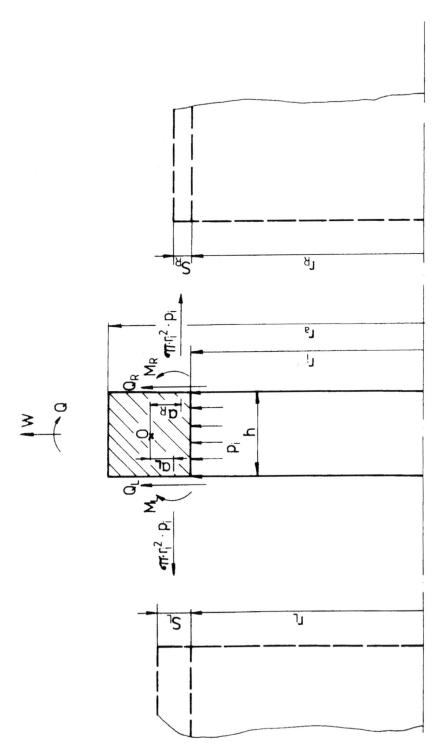

Fig. 2.6 Ring.

The meaning of the abbreviations is as follows:

$$D_R = \frac{h^3}{12} \ln\left(\frac{r_a}{r_i}\right)$$

$$a_L = \frac{(r_a + r_i)}{2} - \left(r_L + \frac{s_L}{2}\right)$$

$$a_R = \frac{(r_a + r_i)}{2} - \left(r_R + \frac{s_R}{2}\right)$$

$$A = \frac{h^2}{8\pi E D_R}$$

$$B = \frac{(r_a + r_i)}{4\pi E (r_a - r_i) \cdot h}$$

$$C = \frac{h}{4\pi E D_R} \tag{2.41}$$

Any standard pressure vessel construction can be described and calculated with the help of the force method using the structural elements listed here. The given equations are valid only if both the load and the components themselves are rotationally symmetrical. A shell branch is not dealt with here. Basically it should be taken into account by introducing flexibility coefficients for the secondary branch.

The rigidly of the transverse flexibility coefficients is obtained from the radial deformation of the secondary shell system based on a unit radial force and the rigidity of the torsion coefficients caused by the torsion of the system, using a unit moment.

2.10 COMBINATION OF INDIVIDUAL ELEMENTS

The individual elements are combined whilst maintaining the equality of the force variable vectors and the overall displacement vector on the assumed separation level. The individual deformation equations, given here for a single element can then be summarised to a linear equation system with a symmetrical elasticity matrix. Here, it is essential to ensure that the signs for the displacements and the torsion are correct.

The procedure to be applied when compiling the overall elasticity matrix of a force method model can be described best by using a simple example.

Fig. 2.7 represents such a simple force method model. It comprises a plate, a short cylindrical shell and a hemispherical shell.

The as yet unknown overall displacement or overall rotation of any element in a particular cross-section is made up of the known deformations of the individual element

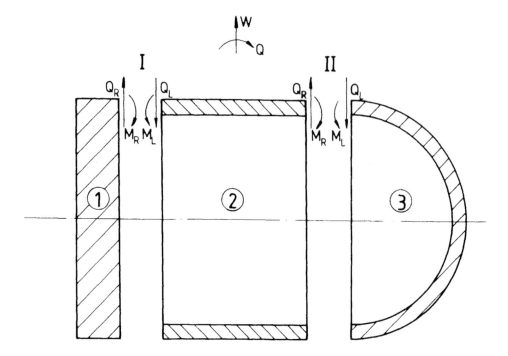

Fig. 2.7 Force method model.

resulting from internal pressure and temperature, and the deformations of the element caused by the unknown boundary loads M and Q. The association with a particular element is indicated by an additional index.

The indices u and b mean: "unknown" and "known".

The three elements in Fig. 2.7 can therefore be described as follows:

$$\begin{Bmatrix} w_R \\ \theta_R \end{Bmatrix}_1^u = \begin{bmatrix} n_{11} & n_{12} \\ n_{21} & n_{22} \end{bmatrix}_1 \begin{Bmatrix} Q_R \\ \theta_R \end{Bmatrix}_1 + \begin{Bmatrix} w_R \\ \theta_R \end{Bmatrix}_1^b$$

$$\begin{Bmatrix} w_L \\ \theta_L \\ w_R \\ \theta_R \end{Bmatrix}_2^u = \begin{bmatrix} n_{11} & n_{12} & n_{13} & n_{14} \\ n_{12} & n_{22} & n_{23} & n_{24} \\ n_{13} & n_{23} & n_{33} & n_{34} \\ n_{14} & n_{24} & n_{34} & n_{44} \end{bmatrix}_2 \begin{Bmatrix} Q_L \\ M_L \\ Q_R \\ M_R \end{Bmatrix}_2 + \begin{Bmatrix} w_L \\ \theta_L \\ w_R \\ \theta_R \end{Bmatrix}_2^b$$

$$\begin{Bmatrix} w_L \\ \theta_L \end{Bmatrix}_3^u = \begin{bmatrix} n_{11} & n_{12} \\ n_{21} & n_{22} \end{bmatrix}_3 \begin{Bmatrix} Q_L \\ M_L \end{Bmatrix}_3 + \begin{Bmatrix} w_L \\ \theta_L \end{Bmatrix}_3^b \qquad (2.42)$$

The compatibility condition in sections I and II is:

Calculation of Structural Stresses Using the Force Method

$$\left\{ \begin{Bmatrix} w_R \\ \theta_R \end{Bmatrix}_1^u \atop \begin{Bmatrix} w_L \\ \theta_L \end{Bmatrix}_3^u \right\} = \left\{ \begin{matrix} w_L \\ \theta_L \\ w_R \\ \theta_R \end{matrix} \right\}_2^u \tag{2.43}$$

and used from (2.42) applies as follows for section I:

$$\begin{bmatrix} n_{11} & n_{12} \\ n_{21} & n_{22} \end{bmatrix}_1 \begin{Bmatrix} Q_R \\ M_R \end{Bmatrix}_1 + \begin{Bmatrix} w_R \\ \theta_R \end{Bmatrix}_1^b$$

$$= \begin{Bmatrix} n_{11} & n_{12} & n_{13} & n_{14} \\ n_{21} & n_{22} & n_{23} & n_{24} \end{Bmatrix}_2 \begin{Bmatrix} Q_L \\ M_L \\ Q_R \\ M_R \end{Bmatrix}_2 + \begin{Bmatrix} w_L \\ \theta_L \end{Bmatrix}_2^b \tag{2.44}$$

and for section II:

$$\begin{bmatrix} n_{11} & n_{12} \\ n_{21} & n_{22} \end{bmatrix}_3 \begin{Bmatrix} Q_L \\ M_L \end{Bmatrix}_3 + \begin{Bmatrix} w_L \\ \theta_L \end{Bmatrix}_3^b$$

$$= \begin{Bmatrix} n_{31} & n_{32} & n_{33} & n_{34} \\ n_{41} & n_{42} & n_{43} & n_{44} \end{Bmatrix}_2 \begin{Bmatrix} Q_L \\ M_L \\ Q_R \\ M_R \end{Bmatrix}_2 + \begin{Bmatrix} w_R \\ \theta_R \end{Bmatrix}_2^b \tag{2.45}$$

The last two equations, therefore, provide the overall equation system for the force method as per Fig. 2.7

$$\begin{bmatrix} (^1n_{11} - {}^2n_{11}) & (^1n_{12} - {}^2n_{12}) & -{}^2n_{13} & -{}^2n_{14} \\ (^1n_{21} - {}^2n_{21}) & (^1n_{22} - {}^2n_{22}) & -{}^2n_{23} & -{}^2n_{24} \\ -{}^2n_{31} & -{}^2n_{32} & (^3n_{11} - {}^2n_{33}) & (^3n_{12} - {}^2n_{34}) \\ -{}^2n_{41} & -{}^2n_{42} & (^3n_{21} - {}^2n_{43}) & (^3n_{22} - {}^2n_{44}) \end{bmatrix} \begin{Bmatrix} {}^1Q_R \\ {}^1M_R \\ {}^3Q_L \\ {}^3M_L \end{Bmatrix} = \begin{Bmatrix} \Delta w_I \\ \Delta \theta_I \\ \Delta w_{II} \\ \Delta \theta_{II} \end{Bmatrix} \tag{2.46}$$

The matrix is the so-called elasticity matrix N; it is symmetrical and can also be simply compiled for a complicated system. The Δw_I to $\Delta \theta_{II}$ express the corresponding deformation differences in the section planes I and II.

$$\Delta w_I = (w_L)_2^b - (w_R)_1^b \; ; \; \Delta \theta_I = (\theta_L)_2^b - (\theta_R)_1^b$$

$$\Delta w_{II} = (w_R)_2^b - (w_L)_3^b \; ; \; \Delta \theta_{II} = (\theta_R)_2^b - (\theta_L)_3^b \tag{2.47}$$

BIBLIOGRAPHY

1. Podhorsky, Stüben, Die Berechnung der Verformung und der Beanspruchung von Flanschen und ihre Optimierung, *VGB Kraftwerkstechnik,* 57 (1977), Heft 10.
2. Podhorsky, M., Dimensionierung des Zylinders unter Berücksichtigung der Randstörspannungen, *Konstruktion,* 26 (1974).
3. Eßlinger, M., Statische Berechnung von Kesselböden, Springer-Verlag, 1952.
4. Timoshenko, S. P., Theory of Plates and Shells, 2, Editions-McGraw-Hill.
5. Timoshenko, S. P., Theory of Elasticity, 3, Editions-McGraw-Hill.
6. Roark and Young, Formulas for Stress and Strain, McGraw-Hill, NY.
7. Flugge, W., Stresses in Shells (2nd ed), Springer-Verlag, Berlin, 1973.
8. Gill, S. S., The Stress Analysis of Pressure Vessel and Pressure Vessel Components, Pergamon, London, 1970.

3
CALCULATION OF TUBESHEETS

3.1 INTRODUCTION

The tubesheet has a particular status amongst the conventional heat transfer components. The load consists of pressure load in the case of U-tube constructions and an additional nonuniform tube load in the case of all straight-tube constructions. In addition to this, the thermal force which depends on the medium and the physical properties of the materials must also be considered in most instances.

When designing tubesheets it is important to take into account the fact that the tubesheet is weakened by the numerous tubeholes which, in the majority of cases, are not symmetrically distributed. The shells attached to the tubesheet determine the so-called restraint of the tubesheet. This boundary condition is very significant as far as deformation and stress distribution are concerned. A simply supported plate has zero edge restraint while a rigidly fixed plate has infinite edge restraint.

A considerable amount of theoretical and practical work has been carried out in connection with the design of tubesheets. Consequently, extensive literature dealing with the problems and the answers is also available.

Today, fatigue analyses are often carried out using the finite element method. This applies in particular to tubesheets which are supported by straight tubes. The perforated section of the plate is normally described using the effective physical variable, e.g., the corrected Poisson's ratio and the corrected modulus of elasticity. This allows the finite element programs introduced in practice to be used and simplifies significantly the idealising of the structure. Detailed studies can then be carried out using the substructure technique.

A great deal of attention has been paid to calculating the equivalent physical variables of tubesheets. A large number of theoretical papers with the accompanying test results has been published showing very good consensus. In most cases, the tests were carried out on photoelastic models. Both British Standard and also the ASME Code provide the design engineer with appropriate graphs of the equivalent physical variables. Differentiation is made between a rectangular and a triangular pitch.

The tubesheets are normally constructed with a constant thickness although as far as stresses and material load are concerned it would be expedient to design and construct them with a variable thickness [1]. Although this would mean less weight, it is important to consider that it would be more difficult to roll the tubes into such tubesheets.

3.2 DESIGN OF THE TUBESHEET ACCORDING TO STANDARDS

The analytical solution of tubesheets loaded rotationally symmetrically and the derivation of the differential equation are typical ways to mathematically resolve a problem relating to a supporting structure. The standards differentiate between a tubesheet with U-tubes and a tubesheet supported by straight tubes. The reason for this is easy to understand. The supported tubesheets are only dealt with in a few standards due to their complexity. The dimensioning of such a tubesheet is set out correctly and in great detail in the TEMA Code. Many other codes, on the other hand, including the German AD-Merkblatt B5, are extremely inaccurate and sometimes incomprehensible, or even incorrect. Extensive discussion would be required to deal with the individual standards and that would go beyond the bounds of this chapter. The majority of the standards quote the same dimensioning formula for the tubesheet with U-tubes except for the calculation coefficient. This formula was derived from the maximum reference stress in the tubesheet [1].

For example the equation for the wall thickness required for round, flat, fully tubed tubesheets with U-tubes is as follows according to AD-Merkblatt B5:

$$h = C D_1 \sqrt{\frac{p \cdot S}{10 \cdot K \cdot v}} \qquad (3.1)$$

The meaning of the formula components is: h, tubesheet thickness; C, calculation coefficient as a function of the boundary conditions (0.35 to 0.4); D_1, tubesheet diameter up to which internal pressure p acts; p, pressure load; S, safety factor according to AD-Merkblatt B0; K, characteristic strength value at design temperature acc. to AD-Merkblatt B0; v, ligament efficiency according to AD-Merkblatt B5.

The wall thickness required for tubesheets which are partially or nonuniformly tubed with U-tubes is calculated using the same formula but with a different design coefficient C. This design coefficient C_4 is determined from a diagram whose logic cannot be verified and seems to be questionable.

3.3 HOW TO CALCULATE THE PERFORATED SECTION

As previously mentioned, despite the work and costs involved, an finite element calculation is carried out for vessels which require a fatigue analysis, and for tubesheet constructions which cannot be allocated to a specific type of standard in order to provide a design analysis of the proposed construction. Such analyses are carried out more and more frequently because with the FE programs being developed and appropriate pre- and postprocessors, the amount of work entailed is decreasing. Programs based on the force method also provide appropriate and inexpensive results [2].

A precise physical description of the materials and, of course, all the dimensions required for this type of deformation analysis with subsequent stress distribution analysis. The tubed section of the tubesheet needs to be treated in a special way.

In most cases, it is not possible to make accurate allowance for the individual boreholes in the tubed section of the tubesheet. Equivalent physical variables E* and μ* have therefore been introduced to take into account the weakening of the tubesheet caused by the boreholes. In the 1950s these values were determined with the help of theoretical and above all photoelastic examinations as a function of the ligament efficiency. The dimensions remain unchanged. The perforated plate is treated as a homogenous plate without any boreholes, but made of a corrected, equivalent material.

Fig. 3.1 shows the correction graphs as given in ASME VIII and BS 5500 for the triangular pitch. A differentiation is made between thin (h ≤ 2t) and thick (h > 2t) tubesheets.

Fig. 3.2 indicates the equivalent physical values for tubesheets with rectangular pitch as per British Standard.

3.4 EXAMPLE OF TUBESHEET ANALYSIS

The tubesheet of a steam generator whose tubes are prestressed and secured in two thin tubesheets is used as an example of tubesheet construction which cannot be dimensioned using the available standards. The tubes are prestressed during manufacturing using SBS (Strain Blocking System). This involves thermal expansion of the tubes shortly before they are secured in the second tubesheet by means of hydraulic expansion. This means that the tubes are appropriately prestressed when the temperature is balanced again.

It is a known fact that two opposite effects occur in systems comprising two tubesheets and straight tubes. To a large extent, thick tubesheets resist the tube loads and this leads to elevated tube stresses. If thin tubesheets are selected, corresponding tubesheet deformations occur that allow the tube stresses to decrease, but, on the other hand, cause the bending stresses in the tubesheet itself to increase to such an extent that they could exceed the admissible limits. It is advisable to carry out an finite element analysis in order to establish the extent to which the construction is subject to stresses. The tubesheet thickness is selected according to the process requirements. During operation the tubesheet load consists of:

- internal pressure on the shell side
- internal pressure on the tube side
- prestressing of tubes
- temperature difference between the mean tube wall temperature and the mean shell wall temperature

A specific example is a steam generator such as the one shown in Fig. 1.31 in which heat from exhaust gases is used to generate steam. The tubesheet load is as follows:

$$\text{tube side} \quad p = 33.5 \text{ bar}$$
$$\text{shell side} \quad p = 120 \text{ bar}$$

The theoretical prestressing of the tubes amounts to 2.3 mm per tubesheet. The temperature difference between the tube and the shell is 60°C. The deformation of the 30 mm thick tubesheet in the four individual load cases and the appropriate superposition is shown in Fig. 1.34. The bending stresses in the critical transition area between the tubesheet and the shell are plotted in Fig. 3.3. The stresses must be broken down into individual components and appropriately taken into account according to the applicable technical regulations. The calculated stresses must be multiplied by the corresponding correction factors in the tubed section of the tubesheet. The reciprocal value of the ligament efficiency v should be used to calculate the membrane stress. In many cases, however, a separate FE model is used as a detail structure. It adopts the moments calculated in the entire structure as well as the transverse and membrane forces at the boundaries.

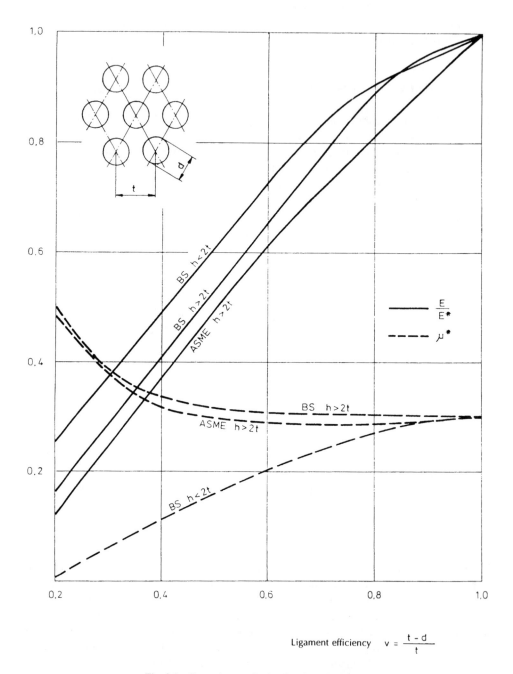

Fig. 3.1 Correction graphs for the triangular pitch.

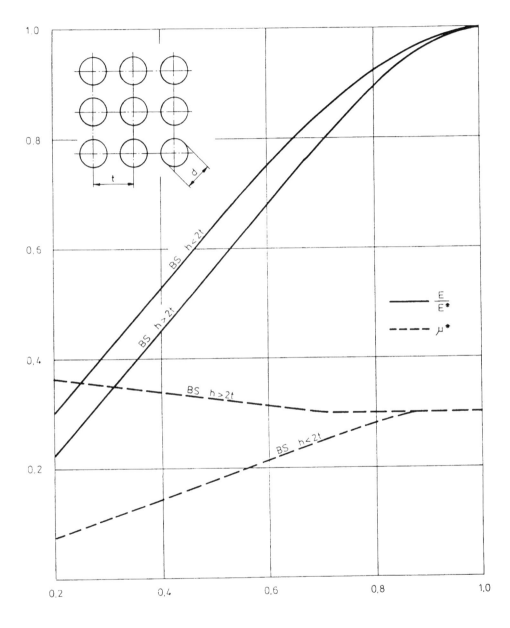

Fig. 3.2 Correction graphs for rectangular pitch.

Ligament efficiency $v = \dfrac{t-d}{t}$

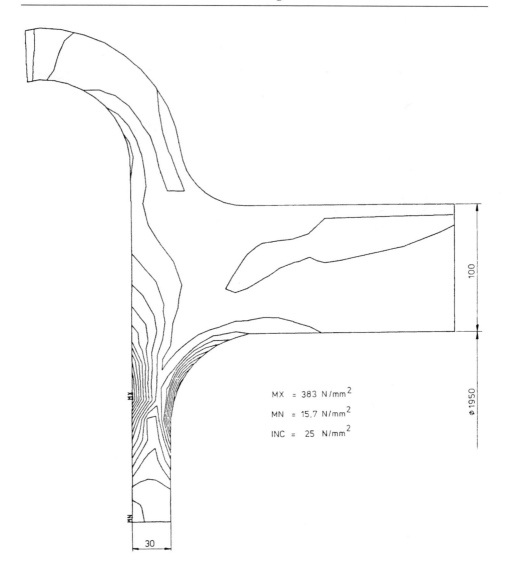

Fig. 3.3 Stress pattern at the tubesheet transition after superimposing load cases 1–4.

BIBLIOGRAPHY

1. Podhorsky, M., Kreisplatten mit veränderlicher Dicke, ihre Berechnung und Vorteile, *Konstruktion,* 25 (1973), Heft 3.
2. Podhorsky, M., Dimensionierung des Zylinders unter Berücksichtigung der Randstörspannungen, *Konstruktion,* 26 (1974), Heft 10.
3. O'Donnell, W.-J. and Langer, B. F., Design of Perforated Plates, *Journal of Engineering for Industry,* August 1962.
4. Laberge, C. A. and Baldur, R., The Effect of Corner Radius on Plate-Cylinder Intersection, *Journal of Pressure Vessel Technology,* February 1990.

5. Gardner, K. A., Heat-Exchanger Tube-Sheet Design, *Trans ASME,* 70 (1948), A-377–385.
6. Gardner, K. A., Heat-Exchanger Tube-Sheet Design-2, Fixed Tube Sheets, *Trans ASME,* 74 (1952), A-159–166.
7. Osweiller, F., Evolution and Synthesis of Effective Elastic Constants for the Design of Tubesheets, *J. Pressure Vessel Technol.,* 111 (1989), 209–217.
8. Galletly, G. D., Optimum Design of Thin Circular Plates on Elastic Foundation, *Proc. Inst. of Mec. Engineers,* 173 (27) (1995), 3–12.
9. Miller, K. A. G., The Design of Tubeplates in Heat Exchangers, *Proc. Inst. of Mech. Engineers,* 18 (1952), 215–231.
10. Soler, A. I. et al., A Proposed ASME Section VIII Division 1, Tubesheet Design Procedure, 1990 ASME Pressure Vessel Piping Conference, Vol. 186, ASME, NY, 3–11.
11. Osweiller, F., Analysis of TEMA Tubesheet Design Rules—Comparison with Up-to-Date Methods, 1986 ASME Pressure Vessel Piping Conference, Vol. 107, ASME, NY, 1–9.
12. McFarlane, R. A., Practical, Accurate Rules for Tubesheet Design, 1990 ASME Pressure Vessel Piping Conference, Vol. 186, ASME, NY, 21–29.

4
DESIGN OF FLANGES FOR PRESSURE VESSELS

4.1 INTRODUCTION

When speaking of flange function or flange design, one should, of course, refer to the whole entity, i.e., flange/bolt/gasket/flange, or cover. Whether a flange joint is tight or not is determined by the four components acting together as well as their design. Therefore, it is not correct to carry out isolated, dissociated calculations of the bolt, the gasket, the flange, or the cover, as is unfortunately the case in the majority of codes. These codes calculate only the stresses in the various sections and these stresses are then compared with the admissible stresses.

The residual surface pressure in the gasket is not, however, considered. The theoretical operability of the flange joint is not calculated. The tightness of the flange joint depends in this case on the experience and skill of the individual fitter who tightens the bolts.

The reason why the codes do not deal with the flange in its entirety can probably be attributed to the discrepancy between a code, which has to provide simple design formulas, and the complicated statically non indeterminate system which must be resolved separately for the individual load cases.

An analysis is a verification of an existing construction which must be improved, at first intuitively and then through iterative processes, to take account of the tightness and material load requirements. It is not, therefore a simple design calculation but a verification.

The main construction features of a flange/flange joint or a flange/cover joint are listed in Table 1. The combination of the various types of construction gives rise to a large number of variants which all behave differently and require different surface pressures at the gasket and different admissible flange face slopes in order to be leaktight.

In all currently applicable codes the flange is considered separately and assumed to be an equi-dimensional flange/flange pair. The deformation phenomenon of a flange/cover joint differs, however, fundamentally from that of a flange/flange joint. Above all the change in the sealing force according to the internal pressure load is completely different, so that different bolt forces must be applied to identical gaskets in the fitted state in order to guarantee the required tightness under operating conditions. Most gasket joints used in pressure vessel construction are flange/cover joints and not flange/flange joints which, however, are frequently used in piping systems.

A new calculation method for the flange joint which is based primarily on the analysis of tightness has been proposed [1]. The stress analysis is only carried out as the final stage of such a calculation. It is, however, essential to ensure beforehand that adequate surface pressure is available during operation using the calculation of deformation. Figure 4.1 shows the calculation logic of this method. The deformation calculation can be carried out using different calculation approaches and methods.

TABLE 4.1 Components of a Flange/Flange Joint or a Flange/Cover Joint

Flange	Bolt	Gasket	Cover
Welding neck flange	Rigid bolt	Indirect force closure	With constant thickness offset
Weld-on flange	Expansion bolt	Metal O-ring	Tubesheet
Detached flange	• with borehole	Sheathed spring wire ring	Dished head
	• without borehole	Spiral-wound gasket	
	Stud bolt		
		Direct force closure	
		Flat gasket	
		• metallic	
		• plastic	
		Rubber/steel gasket	
		Radial gasket	
		Spiral-wound gasket	
		Grooved gasket	
		Sheathed gasket	
		O-ring gasket	
		Ring joint gasket	
		Lenticular gasket	

Reference is frequently made to the applicable codes when carrying out the preliminary design of the individual components. Then the bolt design, the selected gasket, the flange design according to the codes and, subsequently, according to the deformation calculation are described and discussed.

The discussion of the codes will be restricted to the German Codes and the ASME Code.

4.2 BOLT DESIGN

The bolt is the easiest element of the joint to calculate. Under tensile load it is regarded as a simple rod. It is easy to determine the bolt extension. TRD 309 and AD-Merkblatt B 7 are almost identical. First of all the bolt force is calculated under operating conditions and then the bolt force is calculated for the installed state. According to AD-B7 the admissible loading of the gasket is also checked. According to TRD 309 the bolt force required for the test pressure is also determined.

The required bolt area and the required bolt diameter for a known quantity of bolts is calculated from the bolt force and the admissible bolt stress is established from the yield strength at a corresponding temperature and with a corresponding safety coefficient.

The ASME Code also proceeds in a similar way. In this case, two bolt forces are calculated, one for the design conditions of the gasket, and the other for the required initial deformation of the gasket. The required bolt area is then determined by comparison with the admissible stress.

After rounding up the actually selected bolt area, the value is incorporated in the flange calculation; this is certainly a logical approach, but unfortunately only found in the ASME Code.

The bolt design specified by all three codes is based on bolt forces in the operating load case or in the installed load case. The bolt force in the operating load case is a force on which the tightness of the joint depends. Unfortunately, it has nothing to do with the actual bolt force under operating conditions, irrespective of how it is defined. The bolt force for the installed load case as defined by TRD 309 is made up of two components, the internal pressure force and the sealing force required during operation which is multiplied by the factor 1.1. If the initial deformation force of the gasket required to ensure that the gasket fits properly against the contact surfaces is greater, then it is the decisive value for dimensioning the bolts.

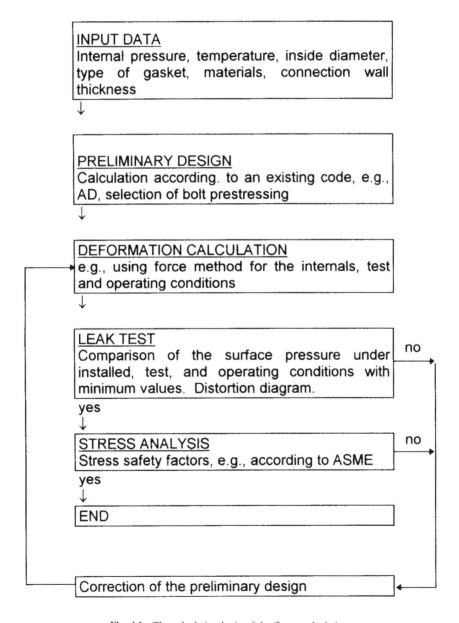

Fig. 4.1 The calculation logic of the flange calculation.

In AD-B 7 and the ASME Code the bolt force in the installed load case is defined as the minimum bolt force and it is identical with the initial deformation force.

Although all codes refer to the bolt force in the installed load case, it is generally acknowledged that the bolts must be tightened with a greater force in order to also ensure a leaktight joint under operating conditions. This is due to the deformation of the construction as a result of internal pressure.

The drop in the bolt force depends on the geometry of the joint, the position and type of the gasket, the operating temperature and a possible temperature difference between the mean temperatures of the bolt and the flange or cover.

4.3 SELECTION OF GASKETS

Various factors influence the selection of the gasket, the most important being the technical requirements with respect to leaktightness, such as operating pressures and temperatures, media, corrosion resistance, operational reliability, diameter, etc. Other criteria also affect the selection of the gasket , e.g., price of the gasket, delivery time, ease of installation, customer requirements, tradition, experience, etc.

The Summary in Table 1 shows that fundamentally there are two different types of gaskets, namely, gaskets with indirect force closure and gaskets with direct force closure.

Gaskets with indirect force closure are pressed into a groove during initial deformation and are not involved in the actual transfer of the sealing force between the flange and the cover or counterflange. They react very sensitively to any clearances between the contact surfaces caused by load displacement at the flange as a result of internal pressure. The advantage is that such a gasket cannot be overloaded because it is enclosed in a groove, (see Fig. 4.2).

The majority of gaskets are installed in direct force closure. The force conditions are then simple and definite. The surface pressure of the gasket is also increased by raising the bolt force in the installed load case. It is relatively easy to calculate the change in the surface pressure and to compare it to the specified minimum surface pressure of the gasket for any operating load cases as a parameter for leaktightness.

For some gaskets the German Codes and the ASME Code specify the so-called sealing coefficients and also indicate at which sealing forces the individual gaskets must be leaktight. They do not, however, indicate, how the actual sealing force is calculated.

A differentiation is made between the sealing force for initial deformation and the operating sealing force. According to the German codes the sealing force for initial deformation is calculated as follows:

$$P_{DV} = \pi . d_D . k_o . K_D \tag{4.1}$$

k_o is a sealing coefficient which can be regarded as being the range of action of the gasket to be used in the calculation. K_D designates the deformation stress of the gasket material. The formulas given in the ASME Code are virtually identical.

The operating sealing force is calculated as follows:

$$P_{DB} = \geq \pi . p . d_D . k_1 . S_D \tag{4.2}$$

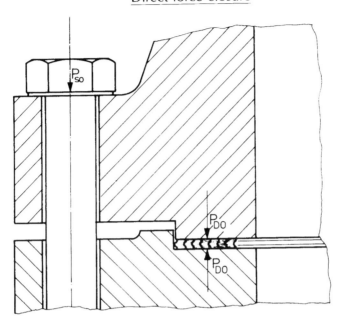

Fig. 4.2 Gasket in direct force closure and indirect force closure.

Sealing coefficient k_1 is the dimension according to a fictitious sealing range. Safety coefficient S_D should take account of uncertain factors under operation. As a minimum, a safety coefficient of 1.2 should be stipulated for the operating case and 1.0 for test pressure. The ASME Code proceeds here in the same way.

Unfortunately, the sealing coefficients specified in the codes are not reliable for higher pressures and temperatures. Furthermore, only a few of the gaskets currently used are mentioned in the codes. When ordering gaskets it is therefore advisable to ask the gasket manufacturer to quote the current sealing coefficients. Some gasket manufacturers indicate the admissible surface pressures and the minimum surface pressures for the installed load case and the operating load case in their catalogues.

These values are normally much higher than the values given in the codes and should be taken into account when dimensioning bolts.

4.4 DIMENSIONING FLANGES ACCORDING TO CODES

Having discussed the bolt calculation and the design of the gasket, one can now turn to the flange itself. There are different types of flange designs. For example, Fig. 4.3 illustrates a welding neck flange, a weld-on flange and a loose ring flange. With all types of flanges, the loading capacity depends, however, on the flexural strength of the flange ring and the flexural strength of the connected shell.

Flange design is dealt with in AD-Merkblatt B 8, in the preliminary standard DIN 2505 dated October 1964, the draft standard DIN 2505 dated November 1972 and in detail in KTA Code 3301.2 dated August 23, 1984. To date the TRD codes have not specified a flange design.

All German codes base the flange calculation on the assumption of a plastic joint which develops at the weakest point.

The required flange resistance W is calculated from the required force in the operating load case and the installed case, using the appropriate lever arm of the force and the allowable stress. The required height of the flange ring can then be specifically calculated for all cross-sections in which a plastic joint can be expected.

The dimensioning formula to be used to calculate the flange height h_F, as specified in AD-Merkblatt B 8, is deduced with the help of Fig. 4.4. The plastic joints are assumed to be either in section A-A or in section B-B.

In order to simplify the derivation, it is assumed that the construction of the conical transition is selected so that section A-A plasticizes first. Then, however

$$A_1 = A_2; \quad e_1 = e_2 = \frac{h_F}{4}$$

$$A_1 = (d_a - d_i - 2d'_L) \frac{h_F}{4} \tag{4.3}$$

The condition of equilibrium of moments acting in a radial plane is:

$$\frac{M_a \, d\alpha}{2\pi} = M_t \, d\alpha + M'_r (d_i + s_F) \frac{d\alpha}{2} \tag{4.4}$$

Design of Flanges for Pressure Vessels

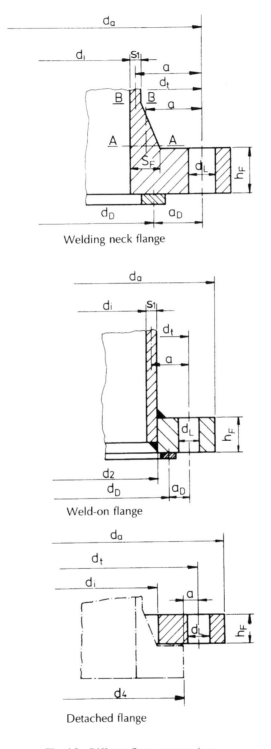

Fig. 4.3 Different flange constructions.

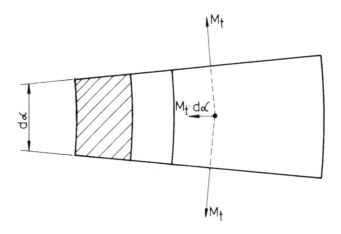

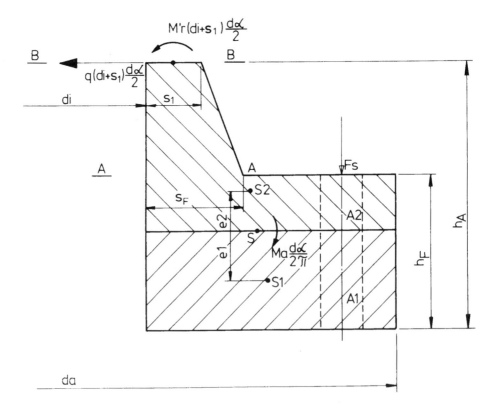

Fig. 4.4 Flange loads according to the codes.

Design of Flanges for Pressure Vessels

The shearing moment is neglected. The individual moments have the following form:
Tangential bending moment

$$M_t = \sigma_t A_1 (e_1 + e_2) = \sigma_t (d_a - d_i - 2d'_L) \frac{h_F^2}{8} \quad (4.5)$$

Bending moment at which the plastic joint occurs:

$$M'_r = \frac{1}{4} \sigma_r (s_F^2 - s_0^2) \quad (4.6)$$

s_0 is the wall thickness required to assume axial force from internal pressure load.
The bending moment:

$$M_a = F_{SB} a \quad (4.7)$$

in which a is the corresponding lever arm of the force according to Fig. 4.3.
For the operating load case the flange resistance W is calculated as follows:

$$W = \frac{F_{SB} \, S}{K} a \quad (4.8)$$

with

$$F_{SB} = \frac{p \pi d_i^2}{40} + \frac{p \pi (d_D^2 - d_i^2)}{40} + \frac{p}{10} \pi d_D S_D K_1 \quad (4.9)$$

In order to simplify the calculation, the installed load case is not considered.
Having applied the individual terms in the physically dominant condition of equilibrium, the equation

$$\sigma_r = \sigma_t = \frac{K}{S}; \quad s_0 = 0; \quad d_a - d_i - 2 \cdot d'_L = b; \quad z = (d_i + s_F) s_F^2 \quad (4.10)$$

is used to obtain the term (6) of AD-Merkblatt B 8 to calculate the required height of the flange ring.

$$h_F = \sqrt{\frac{1.27 \; W - Z}{b}} \quad (4.11)$$

Similarly, the other AD-Merkblatt formulae can be derived to calculate the height of the flange ring, assuming a plastic joint in section B-B or for another type of flange. The expression under the square root can, of course, result in a negative value. This will always be the case when

$$\frac{p}{10} < \frac{(d_i + s_F)s_F}{a(d_D^2 + 4d_D S_D k_1)} \tag{4.12}$$

in the operating load case.

The negative expression under the root sign means that bending moment M_r' in the plastic joint alone is sufficient to assume the value of the extreme moment M_a and therefore, mathematically speaking, an imaginary height is obtained for the flange ring, which is, of course, of no significance. This frequently happens when the internal pressure is low or when there is a thick-walled transition between the flange ring and the cylindrical shell.

Practical experience with heat transfer equipment has shown that flanges with a nominal diameter of more than approximately 1000 mm and designed according to AD-Merkblatt B 8 or to the preliminary standard DIN 2505 have a tendency to leak. The reason for this is that the actual bolt cross-section and therefore the maximum moment in the installed state is much greater than the minimum value used in the formula. Consequently the bolt force F_{SB} does not correspond to the actual bolt force in the operating case; it is actually much greater and the flange ring is caused to tilt to such an extent by the plastic joint that a proper seal cannot be guaranteed.

AD-Merkblatt B8 recommends "that the flange ring angle of inclination ϕ should be restricted to approx. 0.5 to 1° when using plastic and metal/plastic gaskets ..." but does not indicate how the flange ring angle of inclination should be determined.

The draft of DIN 2505 dated November 1972 indicates a formula for the flange ring angle of inclination but at the same time states in a footnote that "the equations for calculating the flange ring angle of inclination have been determined on the basis of numerous tests. It remains to be seen whether they are proven in practice." In conclusion, it can be said that it is only expedient to calculate the flange ring angle of inclination if the actual loads and not, as specified in the German Codes, the required loads are used for the calculation. It is certainly not elegant to specify empirical formulas and then compare them to values applicable in practice. In any case, this method is doomed to failure because of the wide range of flange constructions used in practice. The inclination of the flange ring can only be properly determined for all load cases if the deformation calculation is carried out as presented.

The construction of the so-called "distortion diagram" is also linked to the deformation calculation. Both the preliminary DIN standards dated October 1964 and the draft of DIN 2505 dated November 1972 and the supplements sheet deal with the subject and specify relationships for the calculation of the individual spring coefficients but the required, and not the actual bolt, forces are used in the calculation which is consequently incorrect.

The flange calculation using the ASME Code determines stresses at the point of maximum stress and specifies limits which may not be exceeded. The stresses are determined from the deformation of the conical transition and of the connected shell with the help of graphically represented secondary variables, taking into account the boundary loads. It is assumed that the linear law of elasticity applies.

The bolt force in the installed state is calculated as the product of the arithmetic mean of the minimum and actual bolt surface and of the allowable bolt stress. The bolt force in the operating state consists of the internal pressure force and of the sealing force required for the individual gasket.

The force method is used to calculate the appropriate cross-section variables. The change in the bolt force during the load change and after the force superposition as a result of the internal pressure load is not calculated. Therefore, assumed bolt loads are again used which cannot exist in the actual flange joint.

It is nevertheless more logical and more reliable to calculate the flange/flange joint according to the ASME Code method than according to the German codes. The ASME calculation method generally provides greater flange heights and is, above all for flanges larger than DN 1000, the superior method.

It is possible to subsequently carry out a fatigue analysis and also a stress analysis of the other flange sections. Neither the ASME Code nor the German codes take into account the widely differing conditions at a flange/cover joint where the relatively large angle of inclination of the cover as a result of the internal pressure, results in a significant reduction in the bolt force and, of course, in the sealing force. It is also necessary to consider the clearance between the cover and the gasket which is of vital significance for the tightness of some gaskets.

The deformation conditions are even more complex in the case of a flange joint consisting of a flange and a tubesheet or of two flanges with a tubesheet clamped between them. Experience must then necessarily be a substitute for a definite tightness analysis.

4.5 DIMENSIONING OF FLANGES USING THE DEFORMATION CALCULATION

The calculation of the deformations and stresses in a flange which can be resolved most easily using the force method by assuming rotational symmetry and linear material behaviour, will be derived and discussed below. Unfortunately, this calculation method can be used only as an iterative method for a design calculation. Perhaps that is the reason why it has not been incorporated in the German standards.

Now that pocket calculators belong to the standard tools of the design engineer, the disadvantage of the iterative calculation method has been eliminated. On the contrary, optimisation calculations can very easily be carried out with various parameters and therefore the correct decision can be made about dimensioning and the selected construction details.

As mentioned above, the flange forms only part of the flange/flange or flange/cover joint. For the flange joint to function properly, i.e., to guarantee a leaktight joint in the individual load cases, it is necessary for all four components, that is the gasket, the bolt, the counterflange or the cover and the flange, to be selected and dimensioned correctly. When determining the deformations and the cross-section variables, the most complicated algorithm applies to the flange because of its geometrical shape. Therefore, we will first turn our attention to a weld neck flange with a conical neck as this type occurs most frequently.

As far as load and geometry are concerned, the flange and the connected cylinder as per Fig. 4.5 are axially symmetrical components. The structure is broken down into simple geometrical structures, in this case a flange ring "F", a conical cylindrical shell "K" and a long cylindrical shell "Z".

The displacement and torsion occurring at the interfaces is recorded for these basic elements. The internal pressure and the still unknown cross-section variables at the periphery are used here as the load. These unknown factors are then calculated from the continuity conditions at the interfaces.

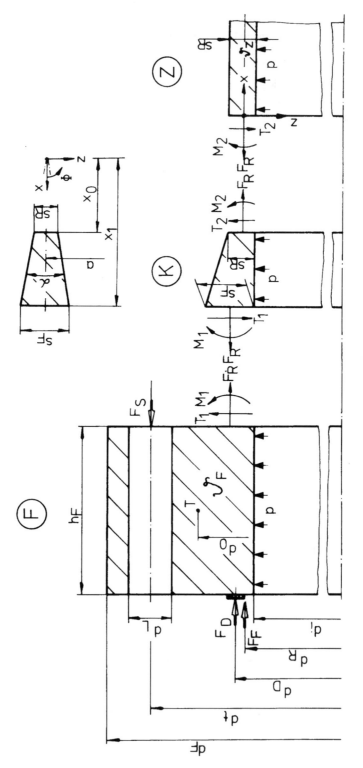

Fig. 4.5 Force calculation method for a flange with a conical neck.

Design of Flanges for Pressure Vessels

Flange ring "F" is loaded in the axial direction by the idealised, uniformly distributed linear bolt force F_s, the sealing force F_D and the circumferential force F_F as well as the tube force F_R. Internal pressure p acts in the radial direction.

The weakening of the flange ring caused by the bolt holes is taken into account as per Ref. 2.

The radius of the centre of rotation is calculated using the terms from Fig. 4.6.

$$r_0 = \frac{r_2 - r_1 - \dfrac{e^2}{t}}{\ln\dfrac{r_2}{r_1} - \dfrac{e}{t}\ln\dfrac{d_t + e}{d_t - e}} \quad (4.13)$$

The resistance to torsion is as follows:

$$D_F = \frac{h_F^3}{12} \ln \frac{d_F}{d_i} \quad (4.14)$$

and the cross-section surface:

$$A_F = A_1 \frac{1}{1 + e/t(A_1/A_2 - 1)}$$

with

$$A_1 = \frac{(d_F - d_i)h_F}{2}$$

and

$$A_2 = e\, h_F \quad (4.15).$$

The torque around the centre of rotation 0 of the flange ring is calculated as follows:

$$M_0 = F_S \frac{(d_t - d_0)}{2} + F_R \frac{(d_0 - d_i - s_F)}{2} + T_1 \frac{h_F}{2} + M_1$$

$$+ F_D \frac{(d_0 - d_D)}{2} + F_F \frac{(d_0 - d_R)}{2} \quad (4.16)$$

and the torsion of the flange ring is then:

$$\Phi_F = \frac{M_0}{2\pi E D_F} \quad (4.17)$$

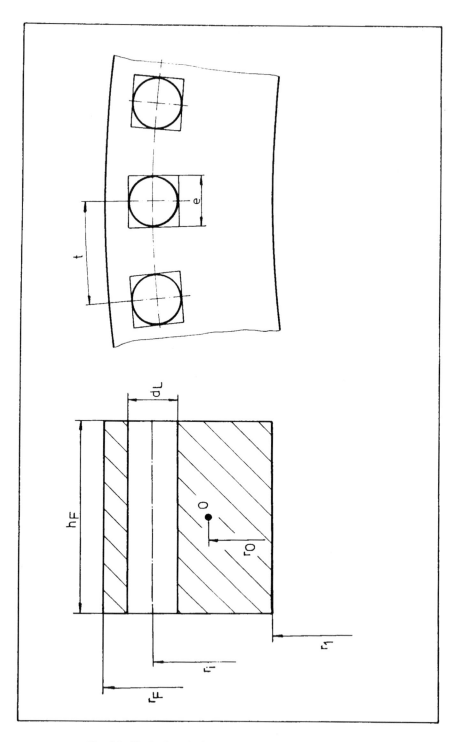

Fig. 4.6 Weakening of the flange ring caused by the bolt holes.

Design of Flanges for Pressure Vessels

The radial displacement of the point of connection to the conical neck is:

$$w_F = -\frac{M_0 h_F}{4\pi E D_F} - \frac{p h_F d_i (d_i + s_F)}{4 E A_F} - \frac{T_i(d_i + s_F)}{4\pi E A_F}$$

$$- (\vartheta_F \alpha_F - \vartheta_z \alpha_z) \frac{(d_i + s_F)}{2} \quad (4.18)$$

The differential equation of the cylindrical shell with the linearly changeable wall thickness as per Fig. 4.5 reads:

$$\frac{d^2}{dx^2}\left(x^2 \frac{d^2 w}{dx^2}\right) + \frac{12(1-\mu^2)}{\alpha^2 a^2} x w = -\frac{12(1-\mu^2)}{E\alpha^3}\left(1 - \frac{\mu}{2}\right) p \quad (4.19)$$

The particular solution is:

$$w_K^P = -\frac{p a^2}{E \alpha x}\left(1 - \frac{\mu}{2}\right) \quad (4.20)$$

and the homogenous solution can, according to [3] be written as:

$$w_K^H = \frac{1}{\sqrt{x}}\left[C_1 \psi_1'(\xi) + C_2 \psi_2'(\xi) + C_3 \psi_3'(\xi) + C_4 \psi_4'(\xi)\right] \quad (4.21)$$

where C_1, C_2, C_3 and C_4 are constants which depend on the boundary conditions.
New secondary variables were introduced

$$\xi = 2\rho\sqrt{x}; \quad \rho = \frac{12(1-\mu^2)}{\alpha^2 a^2} \quad (4.22)$$

in which the mean radius of the conical cylindrical shell a is defined as

$$a = \frac{d_i}{2} + \frac{s_F + s_R}{4} \quad (4.23)$$

The functions $\psi_1'(\xi)$, $\psi_2'(\xi)$, $\psi_3'(\xi)$, and $\psi_4'(\xi)$ are the first derivations according to the argument of the functions

$$\psi_1(\xi) = 1 - \frac{\xi^4}{(2 \cdot 4)^2} - \frac{\xi^8}{(2 \cdot 4 \cdot 6 \cdot 8)^2} - \ldots$$

$$\psi_2(\xi) = -\frac{\xi^2}{2^2} + \frac{\xi^6}{(2 \cdot 4 \cdot 6)^2} - \frac{\xi^{10}}{(2 \cdot 4 \cdot 6 \cdot 8 \cdot 10)^2} + \ldots$$

$$\psi_3(\xi) = \frac{1}{2}\psi_1(\xi) - \frac{2}{\pi}\left[R_1 + \log\frac{\beta\xi}{2}\,\psi_2(\xi)\right]$$

$$\psi_4(\xi) = \frac{1}{2}\psi_2(\xi) + \frac{2}{\pi}\left[R_2 + \log\frac{\beta\xi}{2}\,\psi_1(\xi)\right] \quad (4.24)$$

in which the terms in the square brackets of the last two equations have the following form:

$$R_1 = \left(\frac{\xi}{2}\right)^2 - \frac{S_{(3)}}{(3\cdot 2)^2}\left(\frac{\xi}{2}\right)^6 + \frac{S_{(5)}}{(5\cdot 4\cdot 3\cdot 2)^2}\left(\frac{\xi}{2}\right)^{10} - \ldots$$

$$R_2 = \frac{S_{(2)}}{2^2}\left(\frac{\xi}{2}\right)^4 - \frac{S_{(4)}}{(4\cdot 3\cdot 2)^2}\left(\frac{\xi}{2}\right)^8 + \frac{S_{(6)}}{(6\cdot 5\cdot 4\cdot 3\cdot 2)^2}\left(\frac{\xi}{2}\right)^{12} - \ldots$$

$$S_{(n)} = 1 + \frac{1}{2} + \frac{1}{3} + \ldots + \frac{1}{n}$$

$$\beta = 3.7776 \quad (4.25)$$

The functions (4.24) and their derivations are plotted in Figs. 4.7 and 4.8 for arguments 0 to 6.

In the case of larger arguments the functions can be described by the following terms with adequate accuracy:

$$\psi_1(\xi) = \frac{1}{\sqrt{2\pi\xi}}\, e^{\xi/\sqrt{2}}\, \cos\left(\frac{\xi}{\sqrt{2}} - \frac{\pi}{8}\right)$$

$$\psi_2(\xi) = -\frac{1}{\sqrt{2\pi\xi}}\, e^{\xi/\sqrt{2}}\, \sin\left(\frac{\xi}{\sqrt{2}} - \frac{\pi}{8}\right)$$

$$\psi_3(\xi) = \sqrt{\frac{2}{\pi\xi}}\, e^{-\xi/\sqrt{2}}\, \sin\left(\frac{\xi}{\sqrt{2}} + \frac{\pi}{8}\right)$$

$$\psi_4(\xi) = -\sqrt{\frac{2}{\pi\xi}}\, e^{-\xi/\sqrt{2}}\, \cos\left(\frac{\xi}{\sqrt{2}} + \frac{\pi}{8}\right)$$

$$\psi_1'(\xi) = \frac{1}{\sqrt{2\pi\xi}}\, e^{\xi/\sqrt{2}}\, \cos\left(\frac{\xi}{\sqrt{2}} + \frac{\pi}{8}\right)$$

$$\psi_2'(\xi) = -\frac{1}{\sqrt{2\pi\xi}}\, e^{\xi/\sqrt{2}}\, \sin\left(\frac{\xi}{\sqrt{2}} + \frac{\pi}{8}\right)$$

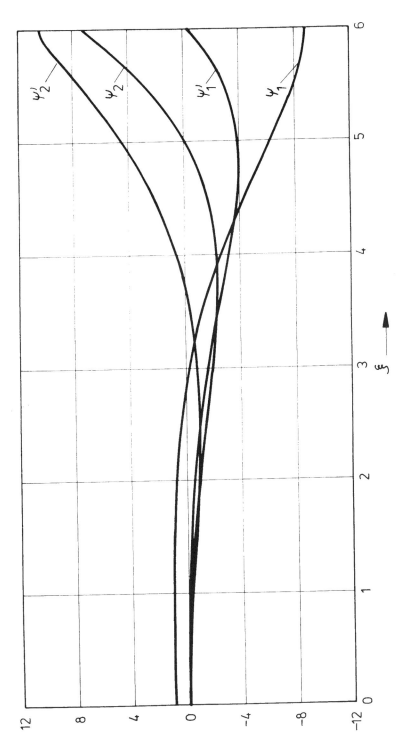

Fig. 4.7 Functions $\psi_1(\xi)$, $\psi_2(\xi)$ and their derivations.

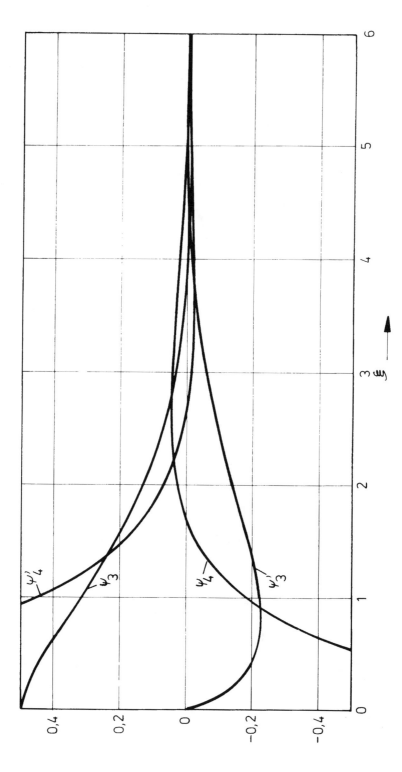

Fig. 4.8 Functions $\psi_3(\xi)$, $\psi_4(\xi)$ and their derivations.

$$\psi_3'(\xi) = -\sqrt{\frac{2}{\pi\xi}}\ e^{-\xi/\sqrt{2}}\ \sin\left(\frac{\xi}{\sqrt{2}} - \frac{\pi}{8}\right)$$

$$\psi_4'(\xi) = \sqrt{\frac{2}{\pi\xi}}\ e^{-\xi/\sqrt{2}}\ \cos\left(\frac{\xi}{\sqrt{2}} - \frac{\pi}{8}\right) \qquad (4.26)$$

The displacement of the short cylindrical shell with variable wall thickness can then be written as :

$$w_K = -\frac{pa^2}{E\alpha x}\left(1-\frac{\mu}{2}\right) + \frac{1}{\sqrt{x}}\left[C_1\psi_1'(\xi) + C_2\psi_2'(\xi) + C_3\psi_3'(\xi) + C_4\psi_4'(\xi)\right] \qquad (4.27)$$

and the torsion as the first derivation with respect to x

$$\begin{aligned}\frac{dw_K}{dx} &= \frac{pa^2}{E\alpha x^2}\left(1-\frac{\mu}{2}\right) + \frac{1}{2x\sqrt{x}}\left\{C_1\left[\xi\psi_2(\xi) - 2\psi_1'(\xi)\right]\right.\\ &\quad - C_2\left[\xi\psi_1(\xi) + 2\psi_2'(\xi)\right] + C_3\left[\xi\psi_4(\xi) - 2\psi_3'(\xi)\right]\\ &\quad \left. - C_4\left[\xi\psi_3(\xi) + 2\psi_4'(\xi)\right]\right\}\end{aligned} \qquad (4.28)$$

The cross-section variables $M_{(x)}$ and $T_{(x)}$ are obtained after further differentiation with respect to x

$$\begin{aligned}M(x) &= -D_K\frac{dw_K^2}{dx^2} = \frac{pa^2\alpha^2(1-\mu/2)}{6(1-\mu^2)} - \frac{E\alpha^3\sqrt{x}}{48(1-\mu^2)}\\ &\quad \cdot\left\{C_1\left[\xi^2\psi_2'(\xi) - 4\xi\psi_2(\xi) + 8\psi_1'(\xi)\right] - C_2\left[\xi^2\psi_1'(\xi) - 4\xi\psi_1(\xi) - 8\psi_2'(\xi)\right]\right.\\ &\quad \left. + C_3\left[\xi^2\psi_4'(\xi) - 4\xi\psi_4(\xi) + 8\psi_3'(\xi)\right] - C_4\left[\xi^2\psi_3'(\xi) - 4\xi\psi_3(\xi) - 8\psi_4'(\xi)\right]\right\}\end{aligned} \qquad (4.29)$$

$$\begin{aligned}T(x) &= \frac{dM(x)}{dx} = \frac{E\alpha^3\rho^2}{24(1-\mu^2)}\sqrt{x}\ \left\{C_1\left[\xi\psi_1(\xi) + 2\psi_2'(\xi)\right]\right.\\ &\quad + C_2\left[\xi\psi_2(\xi) - 2\psi_1'(\xi)\right] + C_3\left[\xi\psi_3(\xi) + 2\psi_4'(\xi)\right]\\ &\quad \left. + C_4\left[\xi\psi_4(\xi) - 2\psi_3'(\xi)\right]\right\}\end{aligned} \qquad (4.30)$$

The deformation and loading of the short conical cylinder is obtained from these equations. The state variables in sections 1 and 2 are obtained by using the variables ξ_o and ξ_1 in the corresponding equations.

The long cylindrical shell with a constant wall thickness is much easier to describe. For example, the radial displacement at the separation point between the cylinder and the conical neck can be calculated according to [4] as follows:

$$w_z = -\frac{p(d_i+s)^2}{4s_R E}\left(1-\frac{\mu}{2}\right) + \left(T_2 \frac{1}{2\beta_z^2 D_z} - M_2 \frac{1}{2\beta_z^2 D_z}\right) \frac{1}{\pi(d_i+s_R)} \tag{4.31}$$

and the torsion

$$\Phi_z = \left(T_2 \frac{1}{2\beta_z^2 D_z} - M_2 \frac{1}{\beta_z D_z}\right) \frac{1}{\pi(d_i+s_R)} \tag{4.32}$$

whereby

$$\beta_z = \frac{\sqrt[4]{3(1-\mu^2)}}{\sqrt{\frac{d_i+s_R}{2} s_R}}$$

and

$$D_z = \frac{s_R^3 E}{12(1-\mu)}.$$

The three elementary structures are brought together using the continuity equation. They assure a constant pattern of displacements, torsion and cross-section variables in the entire flange structure. The continuity equations are at the same time conditional equations for the statically indeterminate boundary forces and moments and also for the four constants C_1 to C_4. The eight equations which form the linear equation system are:

$$w_F = (w_K)_{x=x_1} \qquad T_1 = T_{x=x_1}$$

$$w_Z = (w_K)_{x=x_0} \qquad T_2 = T_{x=x_0}$$

$$\Phi_F = \left(\frac{dw_K}{dx}\right)_{x=x_1} \qquad M_1 = M_{x=x_1}$$

$$\Phi_Z = -\left(\frac{dw_K}{dx}\right)_{x=x_0} \qquad M_2 = M_{x=x_0} \tag{4.33}$$

After solving the linear equation system it is also possible to calculate section variables and deformations at any point of the structure. It is not possible to indicate the coefficients for the individual matrixes in the space available here. It is, however, easy to derive them using the given equations.

Figure 4.9 shows a typical pattern of the radial displace, ent, the lateral force, the axial moment and the axial stress along a conical flange neck and a cylindrical shell. In the case the maximum stress occurs at the beginning of the cylindrical shell.

The question which interests the design engineer most is that of the necessary flange height. It is obvious that the flange height is determined mainly by the allowable stress to which the flange material can be subjected. The minimum height must then be determined for a given material.

Figure 4.10 shows the maximum bending stress as a function of the flange height. The calculation method presented here agrees well with the flange calculation method in ASME VIII. Both methods are based on the calculation of the deformation. The flange geometry and load being studied correspond to Fig. 4.9. This also applies for the following two Figures in which the maximum bending stress in the axial direction is studied as a function of the neck height and neck wall thickness s_F. Compliance with the results obtained using ASME VIII is also very good in this case. This shows that any flange can be optimised by iterative calculation using a suitable program. As illustrated this optimisation can be carried out using various geometric parameters but for economic, functional or constructive reasons it is in most cases not relevant. The axial stresses indicated in this case show that for the flange geometry in question the optimal neck height is approximately 130 mm and that the optimal neck wall thickness is approximately 45 mm.

The most important variation in practice is the mean gasket diameter because it is virtually always possible to change this within limits for most flange joints. The mean gasket diameter is the link between the dimensioning of the bolts and the dimensioning of the flange when the gasket has been selected. Both the bolt force and therefore also the total bolt cross-section as well as the moment of the load of the flange, and thus the flange height are dependent on the mean gasket diameter.

The German and American codes agree as far as this reciprocal influence is concerned. It is, of course, possible to use the two codes together; for example, calculation of the bolts according to the German code and of the flange according to ASME-VIII. In this case, one obtains a flange joint in which the flange is subject to relatively little torsion but the bolts have a greater bolt elongation and are therefore able to absorb, for example, settling of the gasket or temperature differences between flange and bolt without any significant drop in the bolt force.

The required optimal mean gasket diameter can be calculated iteratively or analytically before starting with the actual flange calculation. In this way the minimum flange load is guaranteed right from the start and with the help of appropriate variations of the geometric flange variables it is possible to achieve optimum utilisation of the material.

Figure 4.13 illustrates the pattern of the load moments using an example with calculation variables according to Fig. 4.9. In this case, the selected PTFE[1] gasket is 20 mm wide and 3 mm thick. The bolts were calculated according to AD-Merkblatt B 7 and subsequently the flange height according to ASME VIII. The extent to which the required flange height h_F depends on the mean gasket diameter d_D is obvious.

---

[1]PTFE-Polytetrafluoräthylen.

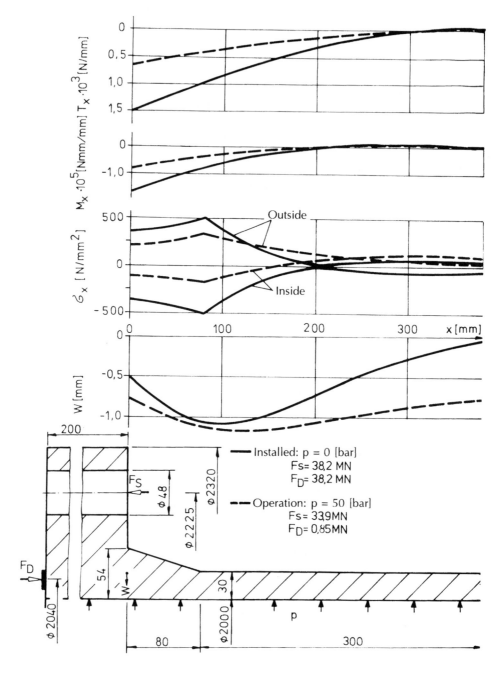

Fig. 4.9 Radial displacement, the lateral force, the axial moment, and the axial stress along a conical flange neck and a cylindrical shell.

Design of Flanges for Pressure Vessels

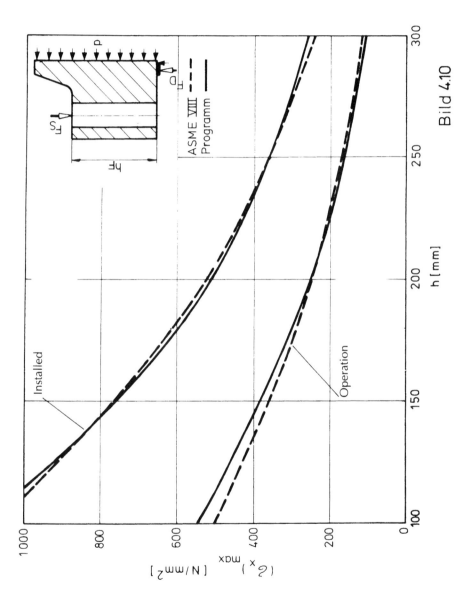

Fig. 4.10 The maximum bending stresses as a function of the flange height h_F.

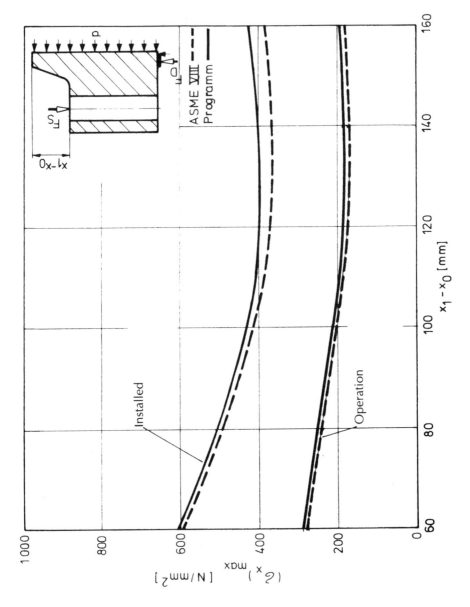

Fig. 4.11 The maximum bending stresses as a function of the neck height (X_1-X_0).

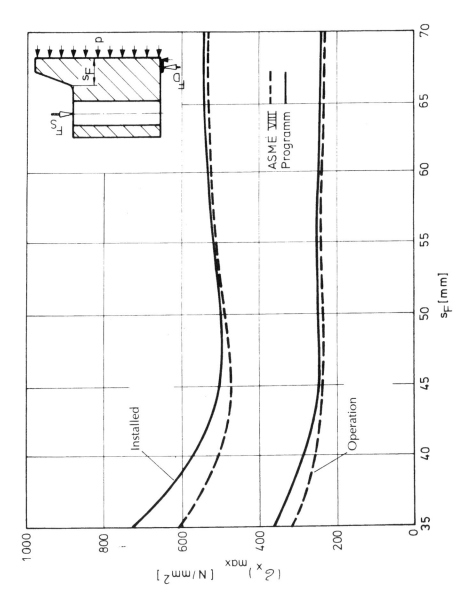

Fig. 4.12 The maximum bending stresses as a function of the neck wall thickness s_F.

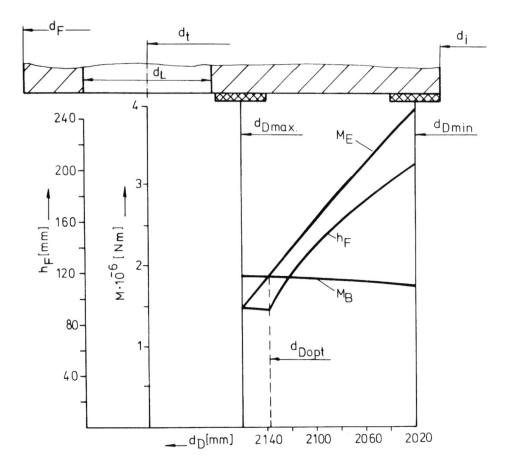

Fig. 4.13 Optimization of the mean gasket diameter.

The equations derived here to calculate the statically indeterminate cross-section variables in a flange with a conical neck show that the description of the short cylindrical shell with linearly variable wall thickness results in complex equations which are difficult to apply in practice. Reference calculations have proved that a simplification in the sense of Fig. 4.14, with a mean wall thickness of the cylindrical shell of $(s_F + s_R)/2$, produces results which are acceptable in practice.

The continuity equation

$$w_F = w_Z \quad \text{and} \quad \Phi_Z = \Phi_F \tag{4.34}$$

and the equations (4.17, 4.18, 4.31 and 4.32) are used to calculate the coefficients for the matrix equation:

$$\begin{bmatrix} a_{11} & a_{12} \\ a_{21} & a_{22} \end{bmatrix} \begin{bmatrix} T_1 \\ M_1 \end{bmatrix} = \begin{bmatrix} b_1 \\ b_2 \end{bmatrix} \tag{4.35}$$

Design of Flanges for Pressure Vessels

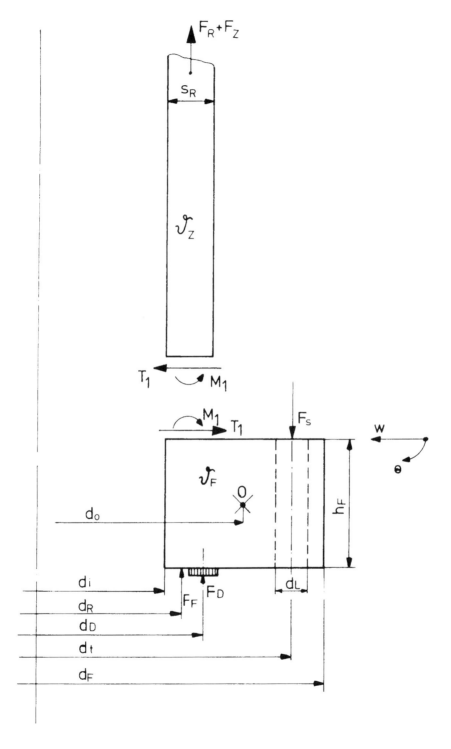

Fig 4.14 Simplified force method for a flange.

as follows:

$$a_{11} = \frac{ED_F}{\beta_z^2 D_z(d_i + s_R)} - \frac{h_F}{2}$$

$$a_{12} = -\frac{2ED_F}{\beta_z D_z(d_i + s_R)} - 1$$

$$a_{21} = -\frac{h_F}{2} - \frac{(d_i + s_R)D_F}{h_F A_F} - \frac{2ED_F}{h_F(d_i + s_R)\beta_z^3 D_z}$$

$$a_{22} = -1 + \frac{2ED_F}{(d_i + s_R)\beta_z^2 D_z h_F}$$

$$b_1 = F_S \frac{(d_t - d_o)}{2} + F_D \frac{(d_o - d_D)}{2} + F_R \frac{(d_o - d_i - s_R)}{2} + F_F \frac{(d_o - s_R)}{2}$$

$$b_2 = F_S \frac{(d_t - d_o)}{2} + F_D \frac{(d_o - d_D)}{2} + F_R \frac{(d_o - d_i - s_R)}{2} + F_F \frac{(d_o - d_R)}{2}$$

$$- p\frac{\pi(d_i - s_R)^2(1 - \mu/2)D_F}{s_R h_F} + p\frac{\pi D_F}{A_F}d_i(d_i + s_R)$$

$$+ \frac{D_F}{h_F} 2\pi E(d_i + s_R)(\vartheta_F \alpha_F - \vartheta_z \alpha_z) \tag{4.36}$$

The coefficients for the righthand side are simplified considerably for the installed state because then:

$$F_S = F_D \qquad F_R = F_F = 0 \tag{4.37}$$

applies.

$$F_D = F_S - F_R - F_F. \tag{4.38}$$

for the operating state.

Having determined the statically indeterminate cross-section variables T_1, M_1, the stresses along the cylindrical shell are calculated and compared with the allowable stresses.

4.6 STRUCTURE OF THE RESTRAINT DIAGRAM

The bolt or sealing force as a function of internal pressure and temperature, which can then be used to control the tightness of the joint under the various load cases, is obtained by linking all components of the flange/flange or flange/cover joint. The graph of the individual forces is frequently referred to as the restraint diagram.

The elastic deflections of the flanges in bolt radius section ΔF_1 and ΔF_2, the elastic bolt elongation ΔS and the resilience of the gasket ΔD are required in order to plot the restraint diagram. For a flange/cover joint it is, of course, necessary to calculate the cover deflection ΔH in the bolt section. The resilience of the individual components in the installed state is calculated as follows:

Flange:

$$C_{F0} = \frac{F_{S0}}{\Delta F_0} = \frac{2F_{S0}}{(d_t - d_D) \operatorname{tg} \Phi_{F0}} \tag{4.39}$$

Bolt: For double-ended bolts:

$$C_{S0} = \frac{\pi n E_{S0}}{4} \left(\frac{d_G^2}{l' + 1.1(l_K - l') + 0.8 d_G} \right)$$

and for reduced shaft bolts:

$$C_{S0} = \frac{\pi n E_{S0}}{l_K} \left(\frac{1}{\dfrac{l_s}{d_s^2 - d_i^2} - \dfrac{0.8 d_G + 1.1(l_K - l_s)}{d_s^2 - d_i^2}} \right) \tag{4.40}$$

The geometric variables are given in Fig. 4.15 and the formulas are taken from [5].

Gasket:
$$C_{D0} = \frac{\pi d_D b_D}{h_D} E_{D0} \tag{4.41}$$

The modules of quasi-elasticity of the gasket can be used for example as given by Ref. [4].

Cover:
$$C_{H0} = \frac{16 \pi d_t}{3(1-\mu) f} \left(\frac{t}{d_D} \right)^3 E_{H0} \tag{4.42}$$

The calculation model of the cover is given in Fig. 4.16. The auxiliary variable f means:

$$f = \left[(1-\mu) + (3+\mu)\beta^2\right]\left(\beta - \frac{1}{\beta}\right) - 2(1-\mu)\left(\beta - \frac{1}{\beta}\right) - 4(1+\mu)\beta \ln \beta$$

$$\beta = \frac{d_t}{d_D} \tag{4.43}$$

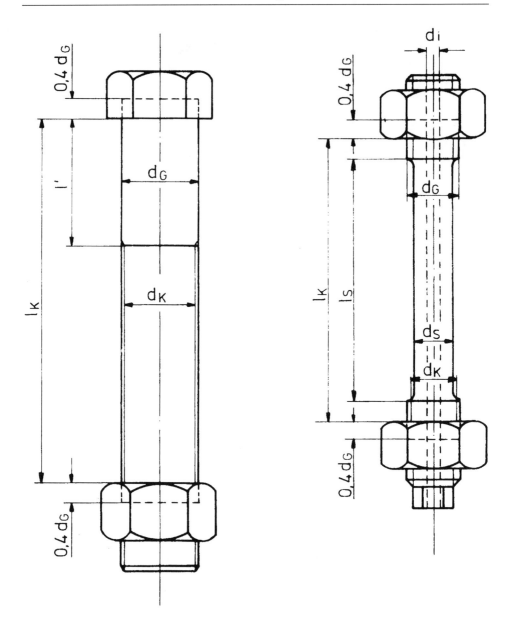

Fig. 4.15 Geometric variables of a double-ended bolt and a reduced shaft bolt.

The flange load and the torsion of the flange change fundamentally when the internal pressure is applied.

The bolt force and the sealing force assume new values. A new load case, start-up, is defined in order to obtain clearly defined conditions. In this case the internal pressure load fully applies to the whereas the temperature does not changed as compared to the installed state. Either the entire static indeterminate system—flange/cover/bolt/gasket— has to be solved (this requires a computer with an appropriate program as will be shown later) or new flange and cover stiffness rates have to be defined in accordance with DIN 2505 (which can be done by manual analytical calculation without a computer) in order to calculate the flange load for this load case, start-up, and to permit the plotting of the restraint diagram.

The formulation according to DIN 2505 is as follows:

Flange:
$$C_{FI} = C_{FO} \frac{M_{00}}{M_{01}} \qquad (4.44)$$

M_{00} is the torsion moment in the installed state

$$M_{00} = F_{S0} \cdot a_D \qquad (4.45)$$

and M_{01} is the torsion moment in the load case 1, i.e., for start-up:

$$M_{01} = F_R \cdot a_R + F_F \cdot a_F + F_D \cdot a_D \qquad (4.46)$$

with

$$F_D = F_{SI} - F_R - F_F$$

and

$$a_R = (d_t - d_i - s_R)\frac{1}{2}$$

$$a_F = (2d_t - d_D - d_i)\frac{1}{4}$$

$$a_D = (d_t - d_D)\frac{1}{2}$$

When inserted in 4.44 that means

$$C_{FI} = C_{FO} \frac{F_{S0}\, a_D}{a_D F_{SI} + \pi/4\left[d_i^{\,2} a_R + \left(d_D^{\,2} - d_i^{\,2}\right)a_F - d_D^{\,2} a_D\right]p} \qquad (4.47)$$

or simplified

$$C_{FI} = C_{FO} \frac{F_{S0}}{F_{SI} + \alpha p} \qquad (4.48)$$

whereby

$$\alpha = \frac{\pi}{4a_D}\left[d_i^2 a_R + \left(d_D^2 - d_i^2\right)a_F - d_D^2 a_D\right]$$

At first the formulation (4.44) seems questionable because the new stiffness rate C_{FI} is a physical variable which is a function of the load and not only of the geometric dimensions and material data as would be expected.

The derivation of formulation (4.44) shows the formula to be correct, and helps to make the equation comprehensible.

Generally, the flange angle of inclination can be described as follows:

$$\Phi_{Fi} = \frac{M_{0i}}{ED_F} \tag{4.49}$$

Stiffness rate C_{Fi} relates to the diameter d_t, i.e.,:

$$C_{Fi} = \frac{2F_{Si}}{(d_t - d_0)\Phi_{Fi}} = \frac{2ED_F}{(d_t - d_0)}\frac{F_{Si}}{M_{0i}} \tag{4.50}$$

The formulation set out in (4.44) follows directly from this formula with $F_{Si} = 1$.

Bolt: $\qquad\qquad\qquad\qquad C_{S1} = C_{S0}$

Gasket: $\qquad\qquad\qquad\quad C_{D1} = C_{D0}$

Cover: $\qquad\qquad\qquad\quad C_{H1} = \dfrac{1}{\dfrac{k_p p}{F_{S1}E_{H0}} + \dfrac{1}{C_{H0}}}$

with $\quad k_p = \dfrac{3(1-\mu^2)}{128}\left(\dfrac{d_t}{t}\right)^3 d_t\beta^2 \left[\dfrac{2(3+\mu) - (1-\mu)\beta^2}{1+\mu}(1-\beta^2) + 6\beta^2 \ln\beta\right] \tag{4.54}$

The stiffness rate for the internal pressure load is obtained from Fig. 4.16 with the formulas given in Ref. [6].

When the flange joint has been subjected to an internal pressure load, the temperature load must also be taken into account in order to obtain the operating conditions. The change in stiffness rate due to temperature is expressed by the change in the physical properties of the material.

Flange: $\qquad\qquad\qquad C_{F2} = C_{F0}\left(\dfrac{F_{S0}}{F_{S2} + \alpha p}\right)\dfrac{E_{F2}}{E_{F0}} \tag{4.55}$

Bolt: $\qquad\qquad\qquad\quad C_{S2} = C_{S0}\dfrac{E_{S2}}{E_{S0}} \tag{4.56}$

Design of Flanges for Pressure Vessels

Gasket:
$$C_{D2} = C_{D0} \frac{E_{D2}}{E_{D0}} \qquad (4.57)$$

Cover:
$$C_{H2} = C_{H1} \frac{E_{H2}}{E_{H0}} \qquad (4.58)$$

Then, we have all the stiffness rates required to plot the restraint diagram which provides the information about the actual bolts forces and stress conditions in the joint during the start-up and operating load cases. The pattern of the sealing force is then known from the condition of equilibrium.

In the installed state the bolts are prestressed with the initial tension force F_{so}. Then, the elastic deflection at the bolt circle is:

$$x_0 = \Delta F_{01} + \Delta F_{02} + \Delta S_0 \qquad (4.59)$$

for a flange/flange joint and

$$x_0 = \Delta F_0 + \Delta H_0 + \Delta S_0 \qquad (4.60)$$

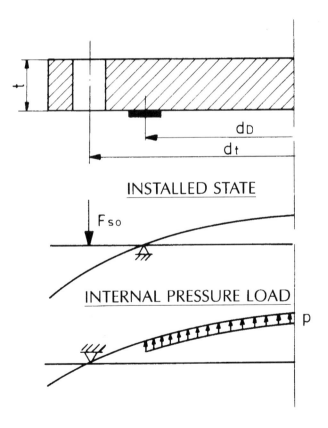

Fig. 4.16 Calculation model for a cover.

for a flange/cover joint. The stiffness rate of the seal loaded with the same force is then ΔD_0. The stiffness rates $C = \text{tg } \alpha$ are then determined as follows:

$$C_0 = \frac{1}{\frac{1}{C_{F01}} + \frac{1}{C_{F02}} + \frac{1}{C_{S0}}} \tag{4.61}$$

or

$$C_0 = \frac{1}{\frac{1}{C_{F0}} + \frac{1}{C_{H0}} + \frac{1}{C_{S0}}}$$

If the internal pressure force caused by internal pressure

$$F_i = F_R + F_F = \frac{\pi}{4} d_D^2 p \tag{4.62}$$

is applied, then the pressure on the seal is relieved to F_{D1} while the bolt force changes from F_{so} to F_{s1}.

This change in the bolt force is decisively influenced by the change in the moment as well as by the elastic behaviour of the joint as a whole. For this reason, angle α_o changes to α_1 in the restraint diagram 4.17 and the conditional equations can now be written.

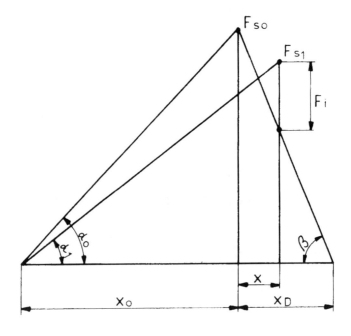

Fig. 4.17 Restraint diagram.

Design of Flanges for Pressure Vessels

For the flange/cover joint

$$\frac{1}{C_1} = \frac{1}{C_{S0}} + \frac{k_p p}{F_{S1} E_{H0}} + \frac{1}{C_{H0}} + \frac{F_{S1} + \alpha p}{C_{F0} F_{S0}} \tag{4.63}$$

and for the flange/flange joint

$$\frac{1}{C_1} = \frac{1}{C_{S0}} + \frac{F_{S1} + \alpha p}{C_{F01} F_{S0}} \left(1 + \frac{C_{F01}}{C_{F02}}\right) \tag{4.64}$$

The new bolt force F_{S1} can now be determined with the help of the geometrical relations from Fig. 4.17.

$$C_{D1} = C_{D0} = C_D = tg\beta = \frac{F_{S1} + F_i}{x_D - x} \tag{4.65}$$

$$C_1 = tg\alpha_1 = \frac{F_{S1}}{x_0 + x} \tag{4.66}$$

$$F_{S1} = \frac{F_{S0} + F_i + C_D x_0}{(1 + C_D / C_1)} \tag{4.67}$$

After inserting the value from (4.63) the new bolt force for a flange/cover joint is

$$F_{S1} = \frac{-b + \sqrt{b^2 - 4ac}}{2a}$$

$$a = \frac{C_D}{F_{S0} C_{F0}}$$

$$b = 1 + C_D \left(\frac{1}{C_{S0}} + \frac{1}{C_{H0}} + \frac{1}{C_{F0}} \frac{\alpha p}{F_{S0}}\right)$$

$$c = C_D \frac{k_p}{E_{H0}} p - F_{S0} - F_i - C_D \frac{F_{S0}}{C_0} \tag{4.68}$$

The bolt force in the start-up load case for the flange/flange joint is calculated analogously to (4.64), the terms a, b, c in (4.68) having the form

$$a = \frac{C_D}{C_{F01} F_{S0}} \left(1 + \frac{C_{F01}}{C_{F02}}\right)$$

$$b = 1 + \frac{C_D}{C_{S0}} + \frac{C_D \alpha p}{C_{F01} F_{S0}}\left(1 + \frac{C_{F01}}{C_{F02}}\right)$$

$$c = -F_{S0} - F_i - C_D \frac{F_{S0}}{C_0} \tag{4.69}$$

Under an operating load, the stiffness rate changes as a function of the modules of elasticity. The bolt force can, in turn, be stated generally, namely with the coefficient (4.77) as

$$F_{S2} = \frac{-b' + \sqrt{b'^2 - 4a'c'}}{2a'}$$

$$a' = \frac{C_{D2}}{F_{S0} C_{F0}} \frac{1}{\varphi}$$

$$b' = 1 + C_{D2}\left(\frac{1}{C_{S0}}\frac{1}{\rho} + \frac{1}{C_{H0}}\frac{1}{\delta} + \frac{1}{C_{F0}}\frac{\alpha p}{F_{S0}}\frac{1}{\varphi}\right)$$

$$c' = C_{D2}\frac{k_p}{E_{H2}}p - \vartheta F_{S0} - F_i - C_{D2}\frac{F_{S0}}{C_0} \tag{4.70}$$

for the flange/cover joint and

$$a' = \frac{C_D}{C_{F01} F_{S0}}\frac{1}{\varphi_1}\left(1 + \frac{C_{F01}}{C_{F02}}\frac{1}{\varphi_2}\varphi_1 \vartheta\right)$$

$$b' = 1 + \frac{C_D}{C_{S0}}\frac{\vartheta}{\rho} + C_D \frac{\alpha p}{C_{F01} F_{S0}}\frac{\vartheta}{\varphi_1}\left(1 + \frac{C_{F01}}{C_{F02}}\frac{\varphi_1}{\varphi_2}\right)$$

$$c' = -\vartheta F_{S0} - F_i - C_D \vartheta \frac{F_{S0}}{C_0} \tag{4.71}$$

for the flange/flange joint. In the coefficients (4.71) index 1 refers to flange 1 and index 2 to flange 2.

Using the equations stated here it is now possible to assess the bolt force and therefore also the bolt load and the sealing force and consequently the tightness of the joint for any load case. It is always possible to calculate the stresses at the most highly stressed points of the flange because the two forces are exerted simultaneously on the flange. This enables the flange to be dimensioned correctly.

The comparison of an elasto-plastic FE-calculation and the calculation method for a flange/cover joint derived here was presented in [1].

It can be said that this method of calculating the new bolt force and sealing force produces higher values if plasticizing is to be anticipated in the seal section.

The advantage of the procedure described here is that the aim is achieved with simple analytical terms.

In the era of desk and pocket calculators a more elegant method is, of course, available which determines the solution factor X_1 directly using matrix calculus. Furthermore, the matrix coefficients for the load case 1 are given here.

The cross-section variables in the installed state are designated as T_{10}, M_{10} and those in load case 1 considered above as T_{11}, M_{11}. The linear equation system for the unknown bolt force F_{S1}, sealing force F_{D1}, boundary force T_{11} and the boundary moment M_{11} can be expressed using the matrix notation, as follows:

$$A\, X_1 = B + A_0\, X_0 \tag{4.72}$$

the matrices A [4,4], B [4,1] and

$$X_0 = \begin{bmatrix} F_{S0} \\ F_{D0} \\ T_{10} \\ M_{10} \end{bmatrix} \tag{4.73}$$

being known. The coefficient matrix A_o [4,4] has the same coefficients as the coefficient matrix A, the only difference being that the factors are

$$\varphi = \delta = \psi = \rho = \vartheta = 1.$$

As previously mentioned, the solution factor is:

$$X_1 = \begin{bmatrix} F_{S1} \\ F_{D1} \\ T_{11} \\ M_{11} \end{bmatrix} \tag{4.74}$$

and the matrix coefficients A [4,4] and B[4.1] can be stated for the flange/cover joint as follows:

$$a_{11} = -1$$
$$a_{12} = 1$$
$$a_{13} = 0$$
$$a_{14} = 0$$

$$a_{21} = -\frac{h_F(d_t - d_0)}{8\pi E_{F0} D_F \varphi}$$

$$a_{22} = -\frac{h_F(d_0 - d_D)}{8\pi E_{F0} D_F \varphi}$$

$$a_{23} = -\frac{1}{2\pi(d_i + s_R)\beta_Z^3 D_{Z0} \psi} - \frac{(d_i + s_R)}{4\pi E_{F0} A_F \varphi} - \frac{h_F^2}{8\pi E_{F0} D_F \varphi}$$

$$a_{24} = \frac{1}{2\pi(d_i + s_R)\beta_Z^2 D_{Z0} \psi} - \frac{h_F}{4\pi E_{F0} D_F \varphi}$$

$$a_{31} = \frac{(d_t - d_0)}{4\pi E_{F0} D_F \varphi}$$

$$a_{32} = \frac{(d_0 - d_D)}{4\pi E_{F0} D_F \varphi}$$

$$a_{33} = -\frac{1}{2\pi(d_i + s_R)\beta_Z^2 D_{Z0} \psi} + \frac{h_F}{4\pi E_{F0} D_F \varphi}$$

$$a_{34} = \frac{1}{\pi(d_i + s_R)\beta_Z D_{Z0} \psi} + \frac{1}{2\pi E_{F0} D_F \varphi}$$

$$a_{41} = \frac{(d_t - d_0)(d_t - d_D)}{8\pi E_{F0} D_F \varphi} + \frac{1}{\rho C_{S0}} + \frac{1}{\delta C_{H0}}$$

$$a_{42} = \frac{(d_0 - d_D)(d_t - d_D)}{8\pi E_{F0} D_F \varphi} + \frac{1}{\vartheta C_{D0}}$$

$$a_{43} = \frac{h_F(d_t - d_D)}{8\pi E_{F0} D_F \varphi}$$

$$a_{44} = \frac{(d_t - d_D)}{4\pi E_{F0} D_F \varphi} \tag{4.75}$$

Design of Flanges for Pressure Vessels

$$b_1 = -\frac{\pi}{4}d_D^2 p - F_Z$$

$$b_2 = \left\{\frac{h_F}{4E_{F0}\varphi}\left[\frac{d_D^2(d_0-d_R) + d_i^2(d_R-d_i-s_R)}{8D_F} + \frac{d_i(d_i+s_R)}{A_F}\right]\right.$$

$$\left. - \frac{(d_i+s_R)^2(1-\mu/2)}{4s_R E_{Z0}\psi}\right\} p + \frac{(d_i+s_R)}{2}(\vartheta_F\alpha_F - \vartheta_Z\alpha_Z) + \frac{h_F(d_0-d_i-s_R)}{4E_{F0}D_F\varphi}F_Z$$

$$b_3 = \frac{d_D^2(d_0-d_R) + d_i^2(d_R-d_i-s_R)}{16E_{F0}D_F\varphi}p - \frac{(d_0-d_i-s_R)}{2\pi E_{F0}D_F\varphi}F_Z$$

$$b_4 = -\frac{(d_t-d_D)\left[d_D^2(d_0-d_R) + d_i^2(d_R-d_i-s_R)\right]}{32E_{F0}D_F\varphi}p - \frac{k_p p}{E_{H0}\delta}$$

$$+h_F\alpha_F\vartheta_F + t\alpha_H\vartheta_H - l_K\alpha_s\vartheta_s - \frac{(d_t-d_D)(d_0-d_i-s_R)}{8\pi E_{F0}D_F\varphi}F_Z \qquad (4.76)$$

with Fz as the additional external force and the coefficients

$$\varphi = \frac{E_F}{E_{F0}}$$

$$\delta = \frac{E_H}{E_{H0}}$$

$$\psi = \frac{E_Z}{E_{Z0}}$$

$$\vartheta = \frac{E_D}{E_{D0}}$$

$$\rho = \frac{E_S}{E_{S0}} \qquad (4.77)$$

The flange/flange joint is also, naturally, described by the matrix equations (4.72). The matrix coefficients are identical with the exception of those indicated below. It is assumed that both flanges have the same geometrical dimensions.

$$a_{41} = \frac{(d_t-d_0)(d_t-d_D)}{4\pi E_{F0}D_F\varphi} + \frac{1}{\rho C_{S0}}$$

$$a_{42} = \frac{(d_o - d_D)(d_t - d_D)}{4\pi E_{F0} D_F \varphi} + \frac{1}{\vartheta C_{D0}}$$

$$a_{43} = \frac{h_F(d_t - d_D)}{4\pi E_{F0} D_F \varphi}$$

$$a_{44} = \frac{(d_t - d_D)}{2\pi E_{F0} D_F \varphi}$$

$$b_4 = -\frac{(d_t - d_D)\left[d_D{}^2(d_o - d_R) + d_i{}^2(d_R - d_i - s_R)\right]}{16 E_{F0} D_F \varphi} p$$

$$+ 2h_F \alpha_F \vartheta_F - l_K \alpha_S \vartheta_S - \frac{(d_t - d_D)(d_o - d_i - s_R)}{4\pi E_{F0} D_F \varphi} F_Z \qquad (4.78)$$

Once the matrix coefficients have been determined, it is easy to calculate the bolt force and sealing force changes under each operating load.

Finally, it can be said that the two methods, i.e., the analytical with the projection (4.44) and the matrix solution according to (4.72) produce comparable results and, therefore, leave the user free to choose which one is to be used in constructing the restraint diagram.

4.7 CAUSES OF POSSIBLE FLANGE LEAKS

Using current computer technology it is possible and expedient to also carry out these calculations for flanges and flange/cover joints which are not very highly exposed. The calculation model reflects the behaviour of a flange very well.

How reliable are the flange joints though as far as the joint tightness is concerned?

Anybody who has dealt with flanges in practice can probably report on leaking flange joints. The leaks frequently occur during intermittent processes such as start-up and shutdown of the plant. In many cases the reason for the leak is very difficult to detect because often several influencing factors are involved. Possible causes are:

- improper assembly
- outdated or incorrect sealing coefficients
- inadequate check of bolt prestressing in the installed state
- the required bolt prestressing is incorrectly determined
- deformation calculation was not carried out, flange or cover gradient is too large
- no strain reserve allowed for intermittent relief of the bolts

Of these six frequent causes of leaks in flange joints, the only one which has not been possible to rectify satisfactorily is the check of the bolt prestressing in the installed state.

The required bolt prestressing can be achieved only by using the following equipment:

- wrench
- torque wrench
- hydraulic prestressing device
- heating elements

Prestressing using the hydraulic equipment is the most elegant and the most accurate method. It is possible to use either a single hydraulic cylinder for each bolt and to increase the force in stages or a single hydraulic unit for all bolts in the flange. This ideal procedure is, however, very expensive and is, therefore, used only for a few flanges, e.g., for the reactor vessel cover.

The method most frequently used to prestress the bolts is that using a torque wrench.

The calculated bolt force is first of all converted to a corresponding torque using the equation

$$M_t = 5 \cdot 10^{-6} F_{S0} d_2 \left[\mu_{tot} \left(1 + \frac{D_K + d_L}{2 d_2} \right) + \frac{P}{\pi d_2} \right] [Nm] \qquad (4.79)$$

d_2 - effective diameter (mm);
D_k - nut across flats (mm);
p - thread pitch (mm)
μ_{tot} - total coefficient of friction (0.14)

and the torque wrench set accordingly. The accuracy of the conversion depends on the actual friction in the thread and is therefore encumbered with a relatively high uncertainty and variance. The inadequacies could be rectified if it were possible to carry out a controlled measurement of the bolt elongation.

The bolt elongation is, of course, a linear function of the bolt tension and the bolt grip l_k. In the case of normal bolts it is between several hundredths and several tenths of a millimetre and can be measured using only relatively complex methods. It is possible to measure the bolt elongation of longer bolts using micrometer screw devices but even then the measuring error is sometimes of the same magnitude as the bolt force portion corresponding to the sealing force under the operating load.

As far as the reliability of a flange joint is concerned, such a controlled measurement is completely inadequate but is accepted because more accurate measuring methods are not available.

It is impossible to calculate the elongation of stud bolts using a micrometer. In this case it is possible, if necessary, to take recourse to expensive measuring equipment which, however, requires a borehole in the centre of the bolt.

Over the past few years, several types of equipment have been introduced to measure the bolt elongation using ultrasound. At first, this method seems ideal but still needs further development due to the physical behaviour of the sound in metals under different stresses. The sound in a non-prestressed bolt is propagated differently to the sound in an identical bolt which has been pre-stressed. This correction of the sound velocity and, therefore, of the bolt length not only depends on the bolt stress but also on the bolt material and must be taken into account.

Currently, the measurement of bolt elongation using ultrasound has unfortunately not yet been adequately solved.

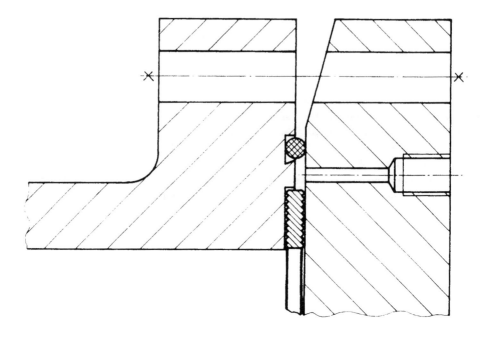

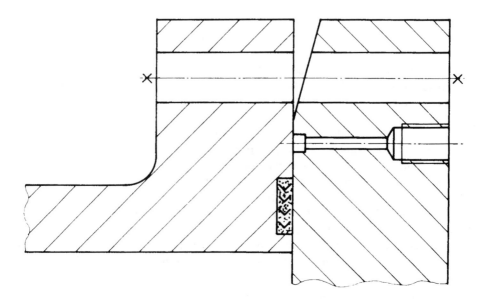

Fig. 4.18 Two possible constructions of a compound seal.

In summary, it can be said that the weakest link in the chain design/manufacture/assembly of a flange/flange or flange/cover joint is the assembly, i.e., the prestressing of the bolts in the installed state.

The reliability could be significantly increased if it were possible to develop a simple, accurate, and reliable method for measuring the bolt elongation.

Finally only one practical question remains to be dealt with, namely: What can be done if a flange starts to leak during operation?

Currently compound sealing processes are being offered on the market. A special compound is injected and allowed to harden to provide temporary tightness and to permit the plant to continue in operation.

Each operator certainly welcomes this new development and the design engineer should, in view of the at times negative experience with flange joints, consider how such a flange joint can be designed so that any leak can be temporarily and easily sealed.

Fig. 4.18 illustrates two possible constructions for such seals, one for a flange joint with direct force closure and one for indirect force closure.

BIBLIOGRAPHY

1. Podhorsky, Vu, Vorschlag eines Berechnungsverfahrens für die Flanschdimensionierung, *VGB KRAFTWERKSTECHNIK*, 64 (1984), Heft 7.
2. Berechnung von Reaktordruckbehältern-Spannungsanalyse nach der Stufenkörpermethode. Fachverband Dampfkessel - Behälter und Rohrleitungsbau E.V., November 1973.
3. Timoshenko, Woinowsky, and Krieger, Theory of Plates and Shells, McGraw-Hill, 1970, NY.
4. Podhorsky, M., Dimensionierung des Zylinders unter Berücksichtigung der Randstörspannungen, *Konstruktion*, 26, Heft 10.
5. Schwaigerer, S., Festigkeitsberechnung im Dampfkesselbehälter und Rohrleitungsbau, Springer-Verlag, 1978.
6. Markus, G., Theorie und Berechnung rotationssymmetrischer Bauwerke, 1967.
7. Waters, E. O., Wesstrom, D. B., Rossheim, D. B., and Williams, F. S. G., Formulas for Stresses in Bolted Flanges Connections, *Trans. ASME,* 59 (1937), 161.
8. Waters, E. O. and Taylor, J. H., The Strength of Piping Flanges, *Trans. Mechanical Eng.,* 49 (1927), 531–542.
9. Lake, G. F. and Boyd, G., Design of Bolted Flanged Joints of Pressure Vessels, *Engs.,* London, 171 (1957), 843.
10. Murray, N. W. and Stuart, D. G., Behaviour of Large Taper Hub Flanges, Proc. Symp. Pressure Vessel Research Towards Better Design, Inc., *Mech. Engs.,* London, 1961, 133.
11. Bickford, J. H., The Bolting Technology Council and the Search for More Accurate Preload, Advances in Bolted Joint Technology, *ASME PVP,* vol. 158.
12. Bickford, J. H., An Introduction to the Design and Behaviour of Bolted Joints, Marcel Dekker, NY.
13. Rossheim, D. B. and Markl, A. R. C., Gasket Loading Constants, Pressure and Piping Design, Collected Papers, 1927–1959, ASME, NY, 1960.

5
METHODS OF FASTENING TUBES IN TUBESHEETS AND HEADERS

5.1 INTRODUCTION

Each heat exchanger in which heat is exchanged in tubes, hence the designation tubular heat exchanger, comprises not only tubes but also a tubesheet or a tube header where the tubes end. Thousands of joints are required for just one heat exchanger and this underlines the importance of and the reason for an easily reproducible, reliable joint of high quality. When manufacturing heat exchangers no other work process is repeated as frequently as that used to fasten the tubes. It is, therefore, very important to select the correct fastening process. A considerable amount of time and expense can be saved in this manner. Fig. 5.1 illustrates the tube bundle of a steam generator which definitely underlines the above facts.

The tube fastening has to meet very stringent requirements. It must be tight, it must be in a position to direct tube forces caused by internal pressure or suppressed thermal expansion into the tubesheet and furthermore it must be corrosion-resistant, especially when stainless steel tubes are involved. The mechanical loads in the joint are mainly periodic and depend on the mode of operation and function of the individual vessel. The corrosive load is not only a function of the medium and the material but also of the residual stresses in the tube. These residual stresses are closely linked to the fastening and manufacturing processes.

The tubes can be fastened in a tubesheet or in a header by

- welding
- roller expansion
- explosion
- expansion

This rough breakdown refers only to the general terms which can be differentiated further. These different methods of fastening are illustrated in Fig. 5.2 in diagrammatic form.

The selection of the proper method of fastening a tube in a heat exchanger not only requires knowledge of the operating conditions of the vessel, the type and properties of the media to be kept separated as well as the tube/tubesheet material combination and the tube/tubesheet geometry but also a good and practical knowledge of the limits, possibilities, advantages and disadvantages of the individual processes.

In the following Sections, an attempt is made to present and describe the individual fastening methods briefly but succinctly. Information in this form should enable the design engineer to make the correct decision about the choice of fastening.

Unfortunately, there is very little literature which deals with this subject. There is no special book or reference work. It is surprising that neither German nor foreign codes pay any attention to this important subject of tube fastening. Despite this, the tube fastenings decide the reliability and availability of entire systems.

Heat Exchangers

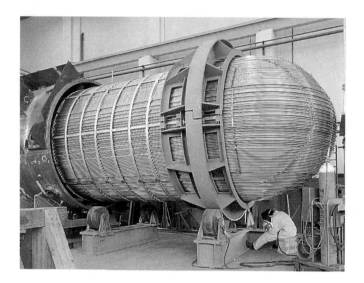

Fig. 5.1 The tube bundle of a steam generator.

Of the German codes only AD-Merkblatt B5 requires adequate safety against extraction in the case of rolled-in tubes or believes to have proved this by complying with allowable stresses of the rolled joint.

The entire procedure set out in the AD-Merkblatt is outdated and completely unsuitable for current methods of fastening. In most cases it is tacitly ignored.

Two totally independent methods of fastening are frequently selected in the case of vessels which are highly significant from a safety point of view, which keep hazardous substances apart or which are essential for the output of the plant as a whole in order to drastically reduce the probability of a leak in the tube/tubesheet joint. In most cases the heat exchanger tubes are welded in and then hydraulically expanded or rolled in. The expansion or rolling and the weld must be independently capable of transferring the loads and, at the same time, of still guaranteeing tightness when the heat exchanger has been in operation for some time. Heat exchangers with several thousands of tubes can operate reliably with this tube fastening redundancy and this fact completely justifies the higher costs involved in fastening the tubes.

5.2 WELDING OF TUBES

Welding heat exchanger tubes into the tubesheets or headers is a modern, reliable and easily reproducible method of tube fastening. It is used mainly for vessels which are subjected to high stress, which keep hazardous media apart or which are highly significant for safety reasons.

In most cases, welding is combined with another method of tube fastening, e.g., with rolling-in or hydraulic expansion.

The tube and tubesheet materials, the pitch as well as the tube diameter and the tube wall thickness then determine the type of weld and weld preparation.

Methods of Fastening Tubes in Tubesheets and Headers

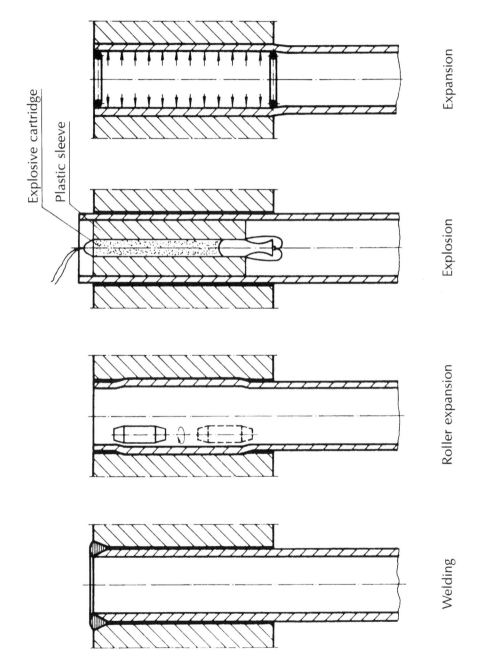

Fig. 5.2 Types of tube/tubesheet joints.

Fig. 5.3 shows some examples of possible weld constructions. A general evaluation of the individual welds cannot be made because the individual advantages and disadvantages can only be assessed in conjunction with the particular application.

Not only must the weld preparation be carried out properly but the weld must be completely clean to ensure the reliable economic success of the tube /tubesheet joint. The large selection of tube/tubesheet welds, the need for uniformity of all joints and the need for freedom from defects has inevitably led to automation of the welding process using automatic tube/tubesheet welding equipment especially designed and developed for this purpose. The main welding parameters are set on this equipment and it controls the rotation of the electrode as well as the supply of inert gas. Normally the welding parameters are tried out and optimised on weld test specimens before actual welding commences. Fig. 5.4 illustrates such a tube/tubesheet weld.

The tubes must be fixed before welding to prevent eccentric weld joints. This is usually done by gentle rolling-in. Rolling-in must be carried out either dry or using water lubrication which causes quite considerable tool wear. Otherwise, even small amounts of residual grease in the weld area would inevitably lead to the formation of pores.

During welding which can be carried out either in the vertical or horizontal position, it is essential to avoid filler metal overhang which would reduce the flow cross-section or cause problems with subsequent rolling-in or expansion.

The problems connected with tube/tubesheet welding relate to welding engineering in general and are not therefore dealt with in greater detail here. If particular tube materials are involved, it is first necessary to carry out extensive preliminary tests which then constitute the manufacturer's wealth of experience in this field. It is therefore understandable that the results of these expensive preliminary tests are considered proprietary and not published in the majority of cases.

5.3 ROLLER EXPANSION OF TUBES

The principle of roller expansion of tubes is very simple and is often compared to the cold-rolling of sheet metal. The tube to be rolled-in can be regarded as an infinite plate whose wall thickness decreases as a result of rolling-out and whose diameter consequently increases until it is identical to the inside diameter of the borehole. The tube then continues to be plastically deformed by further rolling and experience has shown that, as a consequence of non clearly definable material deformation, a certain compressive stress is generated between the tube and the borehole and this secures the tube in the borehole and seals the joint. Fig. 5.5 shows the rolling-in equipment in use.

This method of tube fastening was the first method to be developed and it is still used up to date. The rolling-in equipment was patented as a tool in 1853 and basically has not changed up to the present day. Three or more rolling cylinders installed in a casing are set in a rotating motion using a conical mandrill and pushed outwards by the axial forward-feed of the mandrill. The axes of the mandrill and the rolling cylinders are not parallel to permit the generation of a rotating rolling cylinder motion.

The torque is transmitted to the rolling cylinders by a driving motor via a cardan shaft and the previously mentioned conical mandrill. The rolling cylinders generate rotating radial forces which roll the tube out. Here, the friction between the mandrill and the rolling cylinders and also between the tube inner surface and the rolling cylinders plays a signifi-

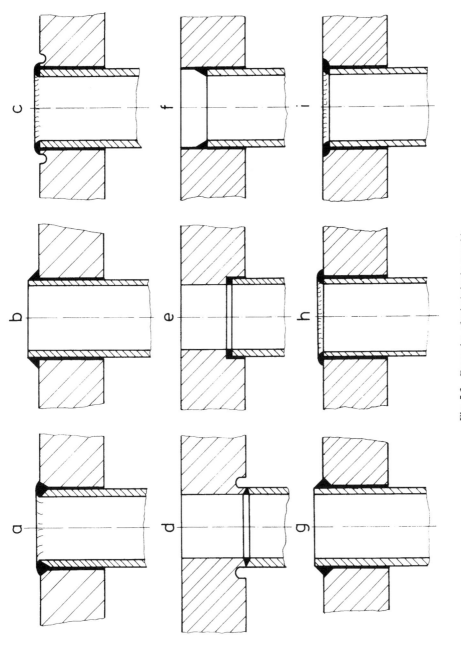

Fig. 5.3 Examples of tube/tubesheet welds.

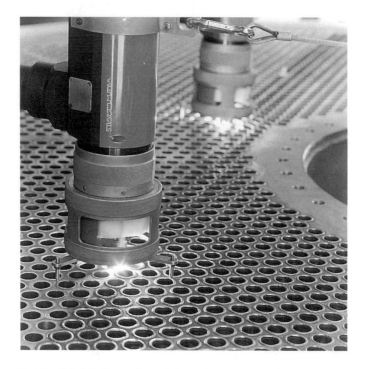

Fig. 5.4 Tube/tubesheet weld carried out with automatic welding equipment.

cant role. The friction converts most of the energy into heat which is why cooling and lubrication are essential when rolling tubes.

A rolling-in controller controls the torque. The controllers available nowadays are electronically controlled and are equipped to set the counterclockwise rotation time and the rest period.

Fig. 5.6 shows a diagram of a complete rolling-in unit. The weight of the roller and the cardan shaft are ideally supported by an expandable suspension. The floor switch facilitates operation.

A normal roller is illustrated in Fig. 5.7. The smallest diameter of the roller is suitable for tubes with a diameter of approx. 10 mm. With smaller diameters it is not possible to transfer the torque because the mandrill cross-section available is not adequate. The rollers are normally 40 mm to 60 mm long. The rolling width which is the effective contact length between the roller and the tube is then 35 mm or 55 mm. The rolling width of rollers with parabola on both sides which are used in particular for stepwise roller expansion, is 5 mm less.

When the tubes are fastened using a combination of welding and rolling, as illustrated in Fig. 5.8, both variants shown there have proved successful. Variant (a) occurs more frequently and the sequence of individual work steps is as follows. First of all, conical rolling is applied at the start of the tube, then the tube is welded into the tubesheet and the weld is subsequently examined by means, for example, of a helium test and the tube is finally expanded.

This procedure has the advantage that the tightness of the weld can be examined before the tube is expanded tightly. The gas bubbles which occur during welding in the weld pool can escape easily. The subsequent expansion must be carried out at an adequate distance from the weld so as not to disturb the previously examined weld. When thick tubesheets are involved the tube is also expanded at the end of the tubesheet in order to avoid possible vibration stress in the tube and to at least partially close the gap in which corrosion products can accumulate as a result of concentration processes. The expansion must permit the thermal expansion of the tube in the axial direction, otherwise all resultant stresses would affect the expanded section or, should this fail, even the weld. In most cases, it is not possible to roll the tube in over the entire thickness of the tubesheet as this would result in an unacceptable tube length increase. Furthermore, this would require several overlapping rolling steps to roll in the tube involving more expenditure and time. The gap which would be enclosed in this way is not desirable because it causes a considerable retardation of the tubesheet temperature compared to the mean tube temperature. However, in many cases it is accepted as a matter of necessity.

Variant (b) of Fig. 5.8 illustrates the following work sequence. First of all, the tube is rolled in, then welded and subsequently expanded, if required. It is not necessary to attach the tube. In this case the weld cannot be examined using the helium test. The gas from any residual gaps must escape forwards through the weld metal which is always problematic.

As already mentioned in the introduction, the rolling-in process is the tube fastening method which has been developed over the longest period of time and with which the most experience has been gained. One would assume that this would facilitate the selection of the correct rolling parameters for a specific case. This is, however, not so.

Fig. 5.5 Rolling-in equipment in use.

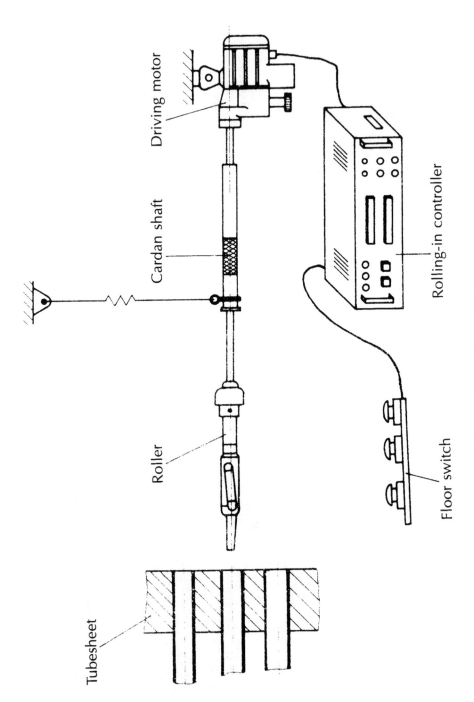

Fig. 5.6 Diagram of rolling-in equipment.

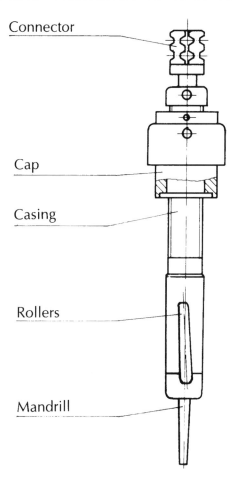

Fig. 5.7 Roller.

In the course of development, a simple parameter has been sought to assess a rolling section. It was then defined as adhesion expansion, also referred to as degree of expansion.

What is adhesion expansion?

Adhesion expansion is the reduction of twice the tube wall thickness expressed as a percentage in relation to the original tube wall thickness s.

Using the terms from Fig. 5.8, the equation for adhesion expansion is as follows:

$$H = \frac{d_{iA} - D_i + 2s}{s} \, 100 \qquad (5.1)$$

and using the term $(D_i - d_{iA}) = 2s_A$ one can write

$$H = 2\left(1 - \frac{s_A}{s}\right) 100 \qquad (5.2)$$

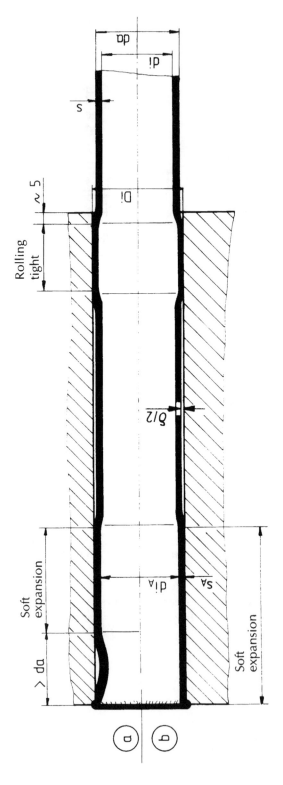

Fig. 5.8 Possible methods for rolling tubes into thick tubesheets.

In some papers, the adhesion expansion is defined as a percentage reduction of the single tube wall thickness. This means that 2s is used in the denominator of equation (5.1) and not the simple wall thickness s. The 2 then disappears before the brackets in (5.2).

The definition (5.1) has gained acceptance but it is still essential to ensure whether the specified value and the calculated value for the adhesion expansion are identical after measurement.

The adhesion expansion according to (5.1) is determined by simply measuring the borehole, the tube inside and outside diameter before rolling and the tube inside diameters afterwards. However, if the tube is observed during the rolling process it can be ascertained that a fictitious adhesion expansion must be calculated at the moment at which the tube just touches the borehole wall and when a firm joint has not yet been formed between the tube and the plate.

The fictitious adhesion expansion H* is derived from the volume parity before and after expansion.

The equation of condition is as follows:

$$\frac{d_i + d_a}{2} s L = (d_a + \delta - s_A) s_A L - \Delta s \frac{d_i + d_a}{2} \qquad (5.3)$$

with tube shortening Δ, rolling length L, and clearance δ.

The solution of the equation of the second degree is

$$s_A = \frac{d_a + \delta - \sqrt{(d_a + \delta)^2 - 4(d_a - s)(1 + \Delta/L)s}}{2} \qquad (5.4)$$

When used in (5.2) it results in the fictitious adhesion expansion

$$H^* = \left\{ 2 - \left[\frac{d_a}{s} + \frac{\delta}{s} - \sqrt{\left(\frac{d_a}{s}\right)^2 + 2\frac{d_a}{s}\frac{\delta}{s} - 4\left(\frac{d_a}{s} - 1\right)\left(1 + \frac{\Delta}{L}\right)} \right] \right\} \qquad (5.5)$$

As a result of the lengthening of the tube during rolling, Δ has a negative sign. In practice this lengthening of the tube can only be measured in very few cases. No significant mistakes can be made if Δ/L is suppressed in relation to 1. The fictitious adhesion expansion depends only on the clearance δ between the tube and the borehole as well as on the outside diameter of the tube and its wall thickness.

The diagrammatic illustration of equation (5.5) is shown in Fig. 5.9. It can be seen that the fictitious adhesion expansion increases greatly as the tube diameter decreases and the clearance increases.

What adhesion expansion should be selected for a given tube/tubesheet material combination? This question has not been clearly answered in the past nor, unfortunately, will be in the future. The reason for this is that there is no mathematical theory on rolling-in and one will not be developed in the future because the process cannot be described mathematically. No well-founded scientific research has yet been carried out into the roller

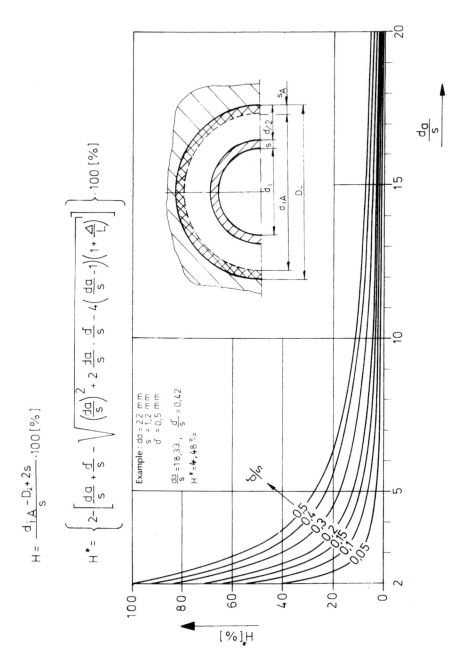

Fig. 5.9 Fictitious adhesion expansion.

expansion process. This is proved by the sparse number of publications on the subject listed in specialized journals. For example, a leading roller manufacturer indicates standard values for adhesion expansion as follows in his catalogue:

Brass and copper tubes	8% to max. 20%
Steel tubes	10% to max. 20%
Steel tubes with high pressure	15% to max. 25%
Steel tubes for high pressure steam boilers with tubes up to 38 mm diameter	15% to max. 25%
Steel tubes with diameter over 38 mm and tube hole ligament stress over 130 N/mm^2	20% to max. 30%
Steel tubes with diameter over 38 mm and tube hole ligament stress over 130 N/mm^2	25% to max. 35%

The following text appears immediately after these "standard values":

"As the conditions vary greatly due to tube dimensions, materials, rolling lengths and operating pressures, it is advisable to carry out several test expansions to establish the correct adhesion expansion."

In other words, the adhesion expansion and therefore also the torque of the rolling equipment must be established by testing. This is what happens in practice. A test does not, however, guarantee that all the tubes of a component are uniformly and adequately fastened because, as already mentioned, the tightness of the joint depends to a high degree on the unreliable physical variable called friction.

Very few tubes and boreholes can be measured on the component to determine the actual expansion and compare it with the specified expansion. The fictitious adhesion expansion is not taken into account at all and this naturally increases the spread of the calculated adhesion expansion even further. In practice, therefore adhesion, expansions must be indicated with tolerances of up to +/– 50% of the rated value, otherwise it would not be possible to produce them in practice. This fact renders the sense of the rolling parameter, adhesion expansion, somewhat doubtful. However there are no other means of comparison.

5.4 HYDRAULIC EXPANSION OF TUBES

The principle of hydraulic expansion is astonishingly simple. The tube is first elastically and then plastically deformed in the expansion section by fluid pressure and pressed on to the borehole wall. By increasing the pressure further the tubesheet or the header are also first elastically and in some cases also plastically deformed. The greater elastic reverse deformation of the plate compared to that of the tube when the pressure is relieved is then decisive for a residual adhesion pressure between the tube and the tubesheet.

The magnitude of the elastic reverse deformation depends only on the geometries and the yield points of the tube and of the plate and it can be calculated arithmetically. The extraction force can be determined from the expansion pressure and the length of the expansion section.

The principle of hydraulic expansion is illustrated in Fig. 5.10. First, the tube is subjected to elastic strain until it starts to flow and the clearance between the tube and the borehole becomes smaller and smaller. If the tube experiences support from the borehole wall, the plate is also subjected to elastic stress and in some cases also to partial plastic stress. If the expansion pressure is removed, reverse deformation occurs to differing degrees because, in most cases, the plate and the tube are made of different materials and have different geometries. Therefore, the lines for elastic deformation have different gradients. For reasons of balance, a radial adhesion pressure p_H then remains between the tube and the borehole. This means that the tube is encircled and thus fastened. This adhesion pressure occurs only when the free reverse deformation of the tubes is less than the free reverse deformation of the tubesheet.

The tube/tubesheet joint can be designed in various ways. Some of them are illustrated in Fig. 5.11. The most frequently used construction (a) is expansion in a smooth borehole. Variant (b) can be used if greater extraction forces are required. The gentle groove can be very simply and economically produced and therefore no significant additional costs are incurred.

In certain special cases it is beneficial to use a joint with an inserted contact sleeve made of plastic or metal. Variant c with one or two enclosed packing elements is suitable only for media with low temperatures and thin tubesheets.

The expansion is carried out using an expansion unit as illustrated in Fig. 5.12. It comprises two separate loops, the oil and the water loop. The oil is pumped from the oil

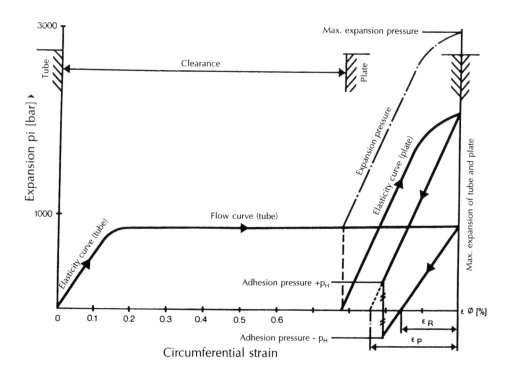

Fig. 5.10 Principle of hydraulic expansion.

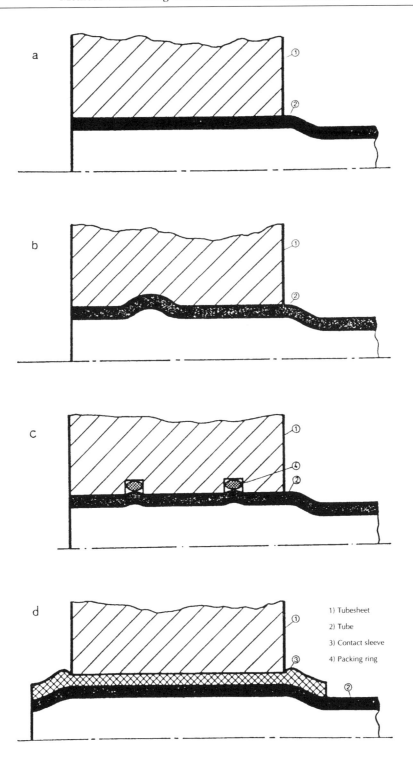

Fig. 5.11 Different types of tube fastening.

1) Tubesheet
2) Tube
3) Contact sleeve
4) Packing ring

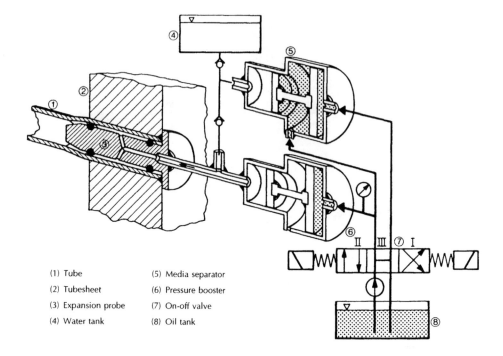

(1) Tube
(2) Tubesheet
(3) Expansion probe
(4) Water tank
(5) Media separator
(6) Pressure booster
(7) On-off valve
(8) Oil tank

Fig. 5.12 Diagram of an expansion unit.

tank to the pressure booster or the media separator. This is controlled by an on/off valve. The function of the media separator is to fill the space between the expansion probe and the tube with water or another suitable fluid.

The required expansion pressure is set using an overflow valve on the oil side. The pressure booster then increases the water pressure by a constant factor. Expansion pistol can be directly equipped with the expansion probe (see Fig. 5.13). The pressure booster generates pressures of up to 4,200 bar and more on the water side, depending on the type of unit. The probe is inserted into the tube and the two sealing elements must then sustain the expansion pressure. The expansion fluid is supplied and discharged through small boreholes between the two sealing elements.

The different types of probes are illustrated in Fig. 5.14. The flexible probe comprises a high- pressure flexible tube which is screwed in between the expansion pistol and the expansion probe. This high pressure flexible tube is available for pressures up to 2,400 bar. It is supple and renders it possible to carry out hydraulic expansion at great depths in tubes which are very difficult to reach. Fig. 5.15 illustrates a possible application.

Tests involving expansions at a depth of 14 m have been successfully carried out. The entire expansion unit and its operation are illustrated in Fig. 5.16. The electronic monitoring system can also be seen on the front face of the unit. This measures the pressure reached for each tube expansion and compares this to the required pressure.

A printer then registers the tube number and the pressure actually reached so that the expansion process can be checked from this record at any time. During the expansion process the pressure is measured by a quartz pressure sensor located directly in the probe.

It is possible to calculate the required expansion pressure because the tube fastening is achieved only as a result of the deformation and expansion of the structure by the fluid pressure. In addition it is possible to determine the stresses in the tubesheet during and after expansion. This is essential for the assessment of the functioning and reliability of a tube/tubesheet joint.

The theoretical verification permits and facilitates the selection of the materials and the geometries of the tubes and of the tubesheet as early as the planning stage.

The theoretical derivation of the hydraulic expansion was published in Ref. [1]. Only the most important terms and equations will be given here.

From a calculation point of view, the tube is an ideal component. The analytical solutions for the tube are known. The internal pressure at which the internal fibres of the tube begin to flow can be calculated in a two-dimensional stress state and using the flow condition according to Misses [1]

$$\sigma_F = \sqrt{3}\sqrt{J_2} \tag{5.6}$$

in which J_2 is the second invariant of the stress deviator, consequently

$$p_i^i = \frac{\sigma_F(u_R^2 - 1)}{\sqrt{3\,u_R^4 + 1}} \tag{5.7}$$

Therein σ_F is the yield point of the tube material and u_R is the ratio of the outside tube diameter to the inside tube diameter.

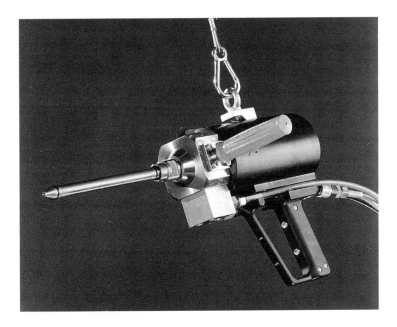

Fig. 5.13 Expansion pistol with expansion probe.

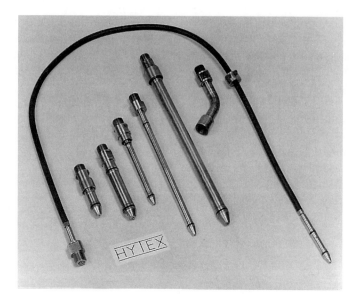

Fig. 5.14 Expansion probes.

If the internal pressure increases further, the plastic zone in the tube extends further outwards. The outer fibres are reached at a pressure of

$$p_i^a = \frac{\sigma_F}{2}\left(u_R^2 - 1\right) \tag{5.8}$$

The radial displacement of a tube is calculated for the elastic region of the material based on stress from the internal pressure p_i

$$w = \frac{p_i r}{E_R} \frac{1+\mu}{u_R^2 - 1}\left[\left(\frac{r_a}{r}\right)^2 + \frac{1-\mu}{1+\mu}\right] \tag{5.9}$$

and on stress from the external pressure p_a

$$w = -\frac{p_a r}{E_R} \frac{1+\mu}{1-1/u_R^2}\left[\left(\frac{r_i}{r}\right)^2 + \frac{1-\mu}{1+\mu}\right] \tag{5.10}$$

The displacement of the outer fibre of the tube is calculated from (5.9) or from (5.10) with $r = r_a$

$$(w)_{r=r_a} = \frac{2 p_i r_a}{E_R\left(u_R^2 - 1\right)} \quad \text{for internal pressure} \tag{5.11}$$

$$(w)_{r=r_a} = -\frac{p_a r_a}{E_R}\left(\frac{(u_R^2+1)}{(u_R^2-1)} - \mu\right) \quad \text{for external pressure}$$

with $p_i = p_i^i$ yields

$$(w)_{r=r_a} = \frac{2\sigma_F r_a}{E_R\sqrt{3u_R^2+1}} \tag{5.12}$$

and with $p_i = p_i^a$ then

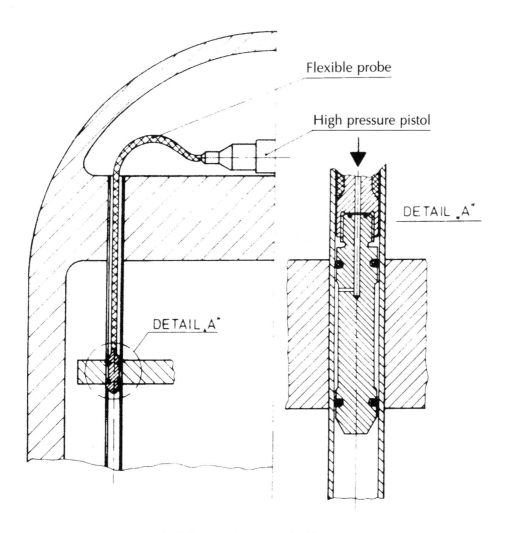

Fig. 5.15 Expansion with the flexible probe.

Fig. 5.16 Hydraulic expansion unit.

$$(w)_{r=r_a} = \frac{\sigma_F r_a}{E_R} \tag{5.13}$$

The calculation model for the tubesheet was first related to a thick-walled tube as shown in Fig. 5.17. In this case all the previously indicated equations also apply to the tubesheet if we substitute E_R by E_p and u_r by u_p. This simplification of the tubesheet geometry is thoroughly acceptable in practice, as for example for the strain gauge measurement, the result of which is shown in Fig. 5.18.

The aim of the expansion process is to build up a residual, sufficiently high adhesion pressure between the tube and the borehole wall which then seals the tube/tubesheet joint and prevents the tube from moving relative to the tubesheet. Therefore

$$\left(w_{r=r_a}\right)_{p_i^i} < \left(w_{R=R_i}\right)_{\left(p_i - p_i^a\right)} \tag{5.14}$$

Methods of Fastening Tubes in Tubesheets and Headers

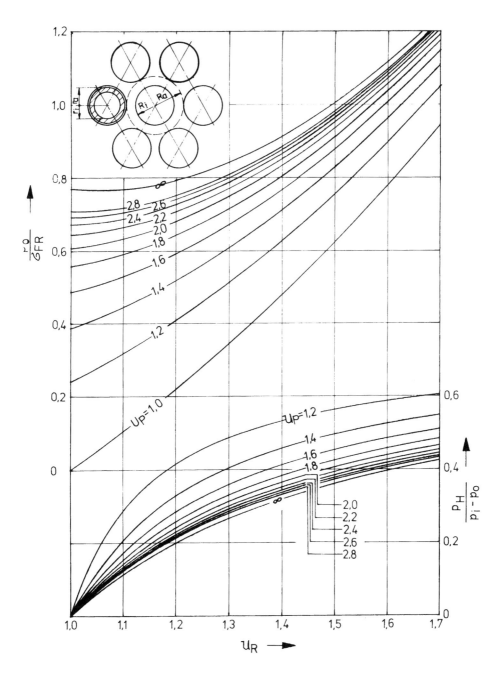

Fig. 5.17 Tube and tubesheet geometry as a function of the limit and adhesion pressures.

is required at the end of the pressure increase.

This inequality results in the condition

$$p_i > \frac{2\sigma_{FR}}{\sqrt{3u_R^4 + 1}} \frac{(u_p^2 - 1)E_p}{[u_p^2(1+l) + 1 - l]E_R} + \frac{\sigma_{FR}(u_R^2 - 1)}{2} \tag{5.15}$$

when it is applied.

The righthand side of the equation (5.15) is referred to as limiting pressure p_o. It represents the expansion pressure p_i at which the reverse deformation of the tubesheet is the same as the reverse deformation of the tube.

In Fig. 5.17 the limiting pressure related to the yield point of the tube is plotted as a function of the radii ratios of the tube/tubesheet joint model assuming $E_R = E_p$. Curve $u_p = 1.0$ indicates the related expansion pressure p_i/σ_{FR} of the tube, curve $u_p = \infty$ the ratios in an infinite tubesheet with a single borehole.

If the tube/tubesheet joint is subjected to a higher expansion pressure than the limiting pressure, then residual adhesion pressure can be anticipated. How high is this adhesion pressure?

The expansion of the tube brought about by the initially unknown adhesion pressure p_H is

$$w_R = -\frac{p_H r_a}{E_R} \frac{1+\mu}{1 - 1/u_R^2} \left(\frac{1}{u_R^2} + \frac{1-\mu}{1+\mu} \right) \tag{5.16}$$

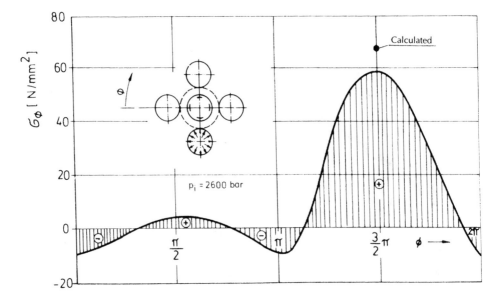

Fig. 5.18 Circumferential stress in the adjacent borehole.

and of the tubesheet

$$w_P = \frac{p_H R_i}{E_P} \frac{1+\mu}{u_P^2 - 1}\left(u_P^2 + \frac{1-\mu}{1+\mu}\right) \tag{5.17}$$

The expansion of the tubesheet model brought about by the pressure fraction $(p_i - p_o)$ is

$$w_P^* = \frac{(p_i - p_0)R_i}{E_P} \frac{1+\mu}{u_P^2 - 1}\left(u_P^2 + \frac{1-\mu}{1+\mu}\right) \tag{5.18}$$

From contact condition

$$w_P^* = w_P - w_R \tag{5.19}$$

one obtains

$$\frac{p_H}{p_i - p_0} = \frac{\beta}{1 + \dfrac{E_P}{E_R} \dfrac{u_P^2 - 1}{u_R^2 - 1} \left[\dfrac{u_R^2(1-\mu) + 1 + \mu}{u_P^2(1+\mu) + 1 - \mu}\right]} \tag{5.20}$$

with $R_i = r_a$ after a simple transformation, i.e., the related adhesion pressure which is illustrated in diagram form in Fig. 5.17, based on the assumption that $E_p = E_R$. The influence of the different module of elasticity must be taken into account in every case when tubes made of titanium, aluminium and copper alloys are to be fastened in tubesheets made of steel. The correction factor β which is included in equation (5.20) corrects the idealised calculation model of the tubesheet and is shown in Fig. 5.19 as a function of the tubesheet geometry. It was determined in a parameter study by comparing the radial displacement at the perforated edge of the simplified tubesheet model and the same displacement of the actual tubesheet calculated using the finite element method. The details can be found in [2].

The extraction force of the tube can then be theoretically calculated as

$$P_R = 2\pi r_a L p_H f \tag{5.21}$$

whereby L is the expansion length and f the friction coefficient between the tube and the borehole wall. This friction coefficient and the inaccuracies of the geometrical shape of the tube and the borehole are responsible for the spreads which should be observed in tube extraction tests.

Below is a model to illustrate how the adhesion pressure of a tube/tubesheet joint is calculated in practice:

Calculation model:

Tubes made of Incoloy 800; $\sigma_{FR} = 475$ N/mm²
Tubesheet made of 22 NiMoCr 37

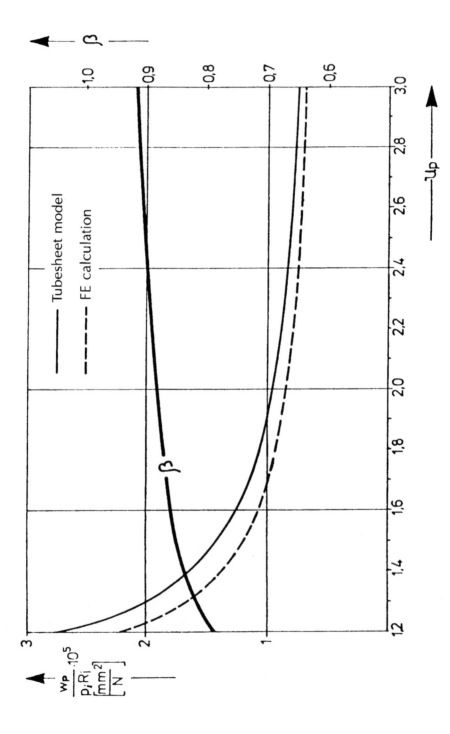

Fig. 5.19 Correction factor β.

Triangular pitch 30 mm
$d_i = 18$ mm
$d_a = 22$ mm
$D_i = 22.3$ mm
$D_a = 37.7$ mm

From Fig. 5.17 we take $u_p = 1.0$ and $u_R = 1.22$

$$\frac{p_0}{\sigma_{FR}} = 0.245$$

The pressure at which the plastic zone has reached the outer fiber of the tube is then

$$p_0 = p_i^a = 0.245 \cdot 475 \cdot 10 = 1163 \text{ bar}$$

For the actual tubesheet geometry we can also take the following from Fig. 5.17

$$U_p = 37.1/22.3 = 1.66$$

$$\frac{p_0}{\sigma_{FR}} = 0.62 \quad \text{and} \quad \frac{p_H}{p_i - p_0} = 0.28$$

The limiting pressure is then calculated as

$$p_0 = 0.62 \cdot 475 \cdot 10 = 2945 \text{ bar}$$

and, for example, for an expansion pressure of 4,200 bar the residual adhesion pressure is

$$p_H = 0.28 \, (4200 - 2945) = 351 \text{ bar}$$

As shown, it is extremely simple to determine the adhesion pressure. The decision as to what the minimum required adhesion pressure is depends on the quality of the surface of the borehole and the tube as well as on the expansion section and the pressure of the medium. A differentiation should be made between fastening and simply expanding the tube. An adhesion pressure of approximately 200 bar is adequate for simple expansion, whereas adhesion pressures of between 300 and 500 bar or more are needed to fasten the tube.

The calculation method described up to now for hydraulically expanded tube/tubesheet joints assumes that the tube will be completely elasticised but that the tubesheet will not be stressed beyond its elasticity limit. This would be desirable but, in practice, for various reasons, material combinations and both tube and tubesheet geometries are chosen, so that residual adhesion cannot be built up without partial plastic deformation of the tubesheet.

This fact also draws attention to the question of the maximum allowable expansion pressure which does not negatively affect the adjacent joints. In such cases, it is now economically viable to carry out an elastoplastic finite element calculation in order to determine the stresses, the pattern, and the extent of the plastic zones during the expansion process.

An example of such an FEM elastoplastic calculation is presented here. The dimensions and the material behaviour can be seen from Fig. 5.20. The shaded area is treated as a substructure. It was assumed that the radial clearance between the tube and the borehole is 0.3 mm. A permanent contact force cannot be built up until this clearance has been reduced to zero through plastic deformation of the tube. In order to realise this condition, the tube was joined to the borehole with a total of seven gap elements.

Fig. 5.21(a) shows the lines of the same reference stresses during the expansion process with p_i = 2,500 bar. The shaded areas are the zones of the tube and the tubesheet which have undergone plastic deformation during the expansion process.

Fig. 5.21(b) shows the stress distribution in the structure when the pressure has been relieved. Similarly Fig. 5.22 shows the lines of the same reference stresses during and after an expansion process with p_i = 3,500 bar. The plastic zone in the tubesheet ligament has extended considerably.

The residual adhesion pressure between the tube and the borehole wall was determined after the pressure had been relieved and compared with the result of the previously mentioned analytical calculation based on equation (5.20).

bar	p_i	p_H	p_i	p_H
Eq. (5.20)	2,500	130	3,500	460
FEM	2,500	141	3,500	464

The results agree very well. Experience gained in practice has confirmed that the values obtained from the elastic analytical calculation can also produce useful results if the tubesheet is partially elasticised.

Fig. 5.23 gives a general picture of the reference stresses during the expansion process in the entire remaining tubesheet structure. The wide "radiation" of the higher expansion pressure of 3,500 bar compared to the lower pressure of 2,500 bar is evident. This illustration gives a certain qualitative assessment of the possible influence on the adjacent tubes which are already fastened, but it is not possible to make a quantitative evaluation of the deformations. Such an evaluation is very important and will shortly be indicated with the help of an elastoplastic parameter study for all tubesheet geometries.

The parameter study was restricted to the usual geometries. The material was assumed to behave ideally elastoplastic without hardening.

Calculations were carried out for the following geometries, t/D_1 = 1.1; 1.2; 1.3; 1.4 using the FE program ANSYS and a structure similar to Fig. 5.20. The start of plastificization and the advance of the plastic zone were observed. For this, relatively small internal pressure increments of Δp = 50 bar were required, each with approximately 100 iterations.

Figs. 5.24 to 5.27 clearly show the spread of the plastic zone for the tubesheet geometries t/D_1 = 1.1 to 1.4.

This is confirmation of the engineering concept, which points to a stress distribution in a thin ligament (t/D_1 = 1.1) corresponding to a square profile stressed to bending by line load. The stress distribution in a thick ligament (t/D_i = 1.4) tends more towards the value

Methods of Fastening Tubes in Tubesheets and Headers

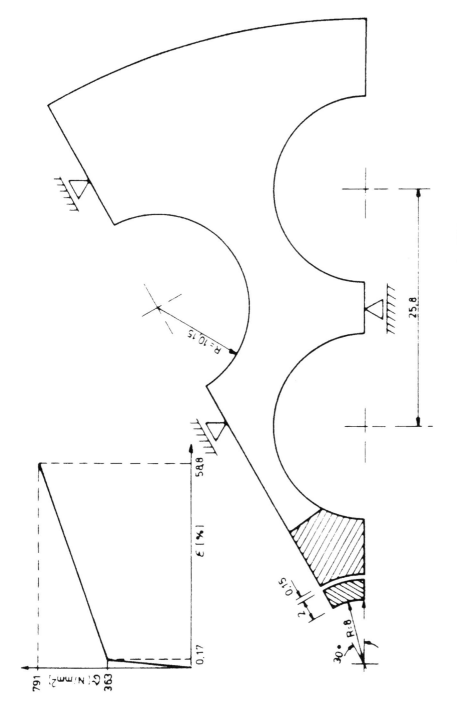

Fig. 5.20 The tube/tubesheet section idealized using FEM.

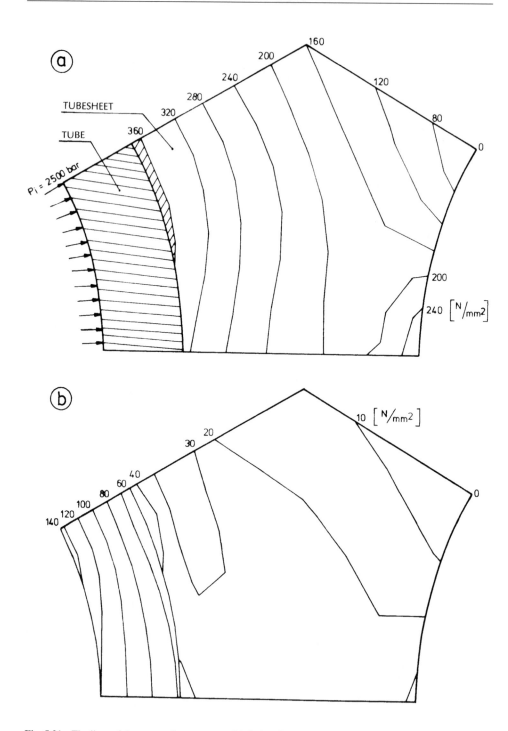

Fig. 5.21 The lines of the same reference stress: (a) during the expansion process with $p_i = 2{,}500$ bar, (b) after the expansion process.

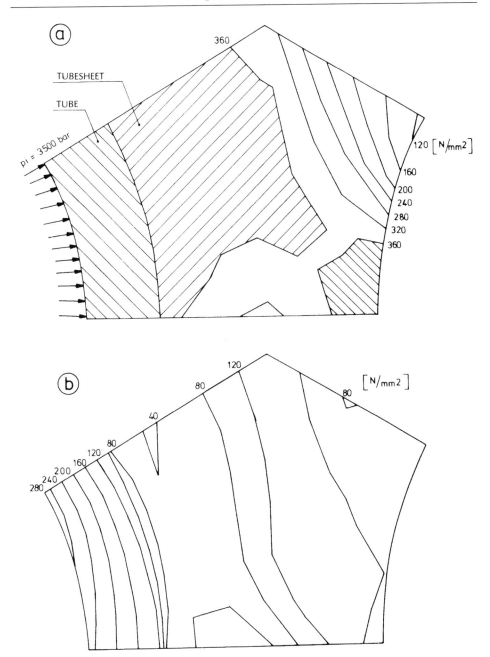

Fig. 5.22 The lines of the reference stress: (a) during expansion with $p_i = 3{,}500$ bar, (b) after expansion.

for a thick-walled tube. In the case of the thick ligament, the plastic zone spreads from the inside outwards, whereas in the case of the thin ligament it spreads from both sides.

It is not easy to say when the load-bearing capacity of the ligament is exhausted. Under no circumstances is it internal pressure which first causes the plastic zones from the inside and from the outside to meet according to Figs. 5.24 to 5.27 because the support effect of the remaining tubesheet and the curvature of the borehole has a great influence on the load-bearing and deformation capacity of the ligament. This observation can also be made from Figs. 5.28 and 5.29.

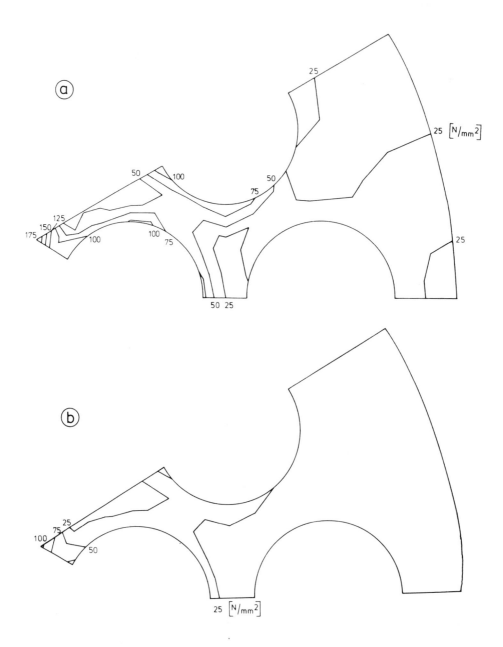

Fig. 5.23 The lines of the same reference stresses in the tubesheet structure: (a) at 3,500 bar expansion pressure, (b) at 2,500 bar expansion pressure.

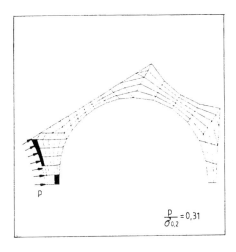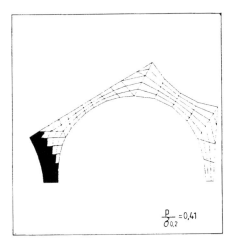
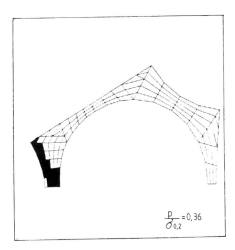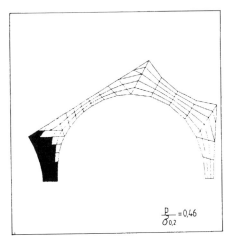

Fig. 5.24 Form of the plastic zone $t/D_i = 1.1$.

The radial displacement of point 1 is plotted in the first illustration, i.e., the indentation of the adjacent borehole as a function of the internal pressure related to the yield point for the four different tubesheet geometries as a parameter. Here, the start of plasticization is indicated by the first symbol of each curve. The plastic zones are penetrated and join at the end of the criss-cross curves.

The broken curves indicate the further loading above this point up to twice the internal pressure of initial plasticization. It is not possible to observe any acceleration of the radial displacement; that means that the load-bearing capacity of the structures up to twice the internal pressure of initial plasticization has still not been reached.

In principle, the same conclusion can be drawn from Fig. 5.29, where the residual reference stress according to Misses is plotted above the relative internal pressure in point 1.

The internal pressure p is defined as follows:

$$p = p_i - p_i^a \quad (5.22)$$

in which p_i represents the actual expansion pressure and p_i^a the internal pressure which causes complete plasticization of the tube according to equation (5.8).

Fig. 5.30 shows the beginning and the end or, more precisely, the penetration of plasticization through the ligament for the different geometries of the tubesheet. The other curves plotted in the diagram are those for double internal pressure relative to initial

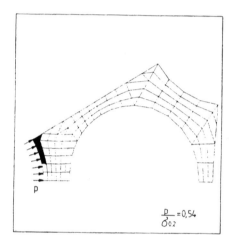
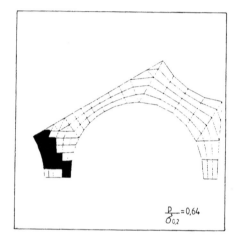
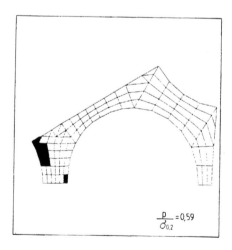
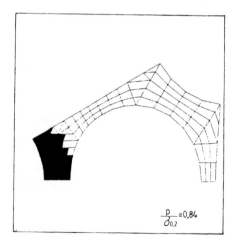

Fig. 5.25 Form of the plastic zone $t/D_i = 1.2$.

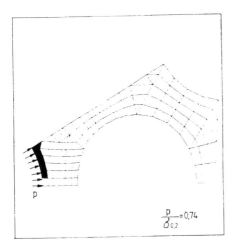 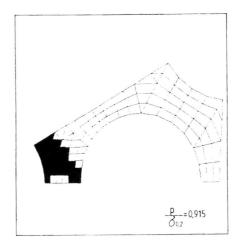
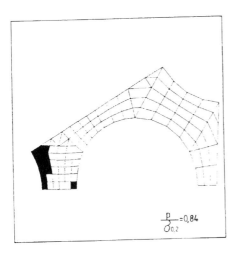 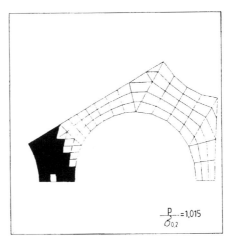

Fig. 5.26 Form of the plastic zone $t/D_i = 1.3$.

plasticization, for the load-bearing capacity of a thick-walled tube as a result of internal pressure and for a square profile stressed to the bending point.

It can be deduced from this study that expansion pressures up to twice the internal pressure at initial plasticization plus the internal pressure required for complete plasticization of the tube may be used when fastening a tube by means of hydraulic expansion, if the deformation of the adjacent borehole is within justifiable limits. This is naturally more often the case with larger t/D_i ratios, i.e., with thicker ligaments than with thin ligaments where the deformation curve is very flat.

Virtually no studies and very little experience are available about the behaviour of a tube/tubesheet joint at high temperatures when the creep properties of the material cause relaxation of the adhesion pressure. Once again the advantage of hydraulic expansion is its calculability. Findings which could be obtained only for other fastening methods from extended tests at high temperatures on samples closely resembling the joint involved, can be determined by calculation for the hydraulic expansion process.

A tube/tubesheet joint produced by applying an expansion pressure of 2,500 bar corresponding to an initial adhesion pressure of 240 bar for more than 2000 hours was calculated assuming following Norton law

$$\dot{\epsilon} = k \cdot \sigma^n \text{ with } k = 1.98 \cdot 10^{-14} \text{ and } n = 7.63 \text{ at } 900°C$$

using an FE-program.

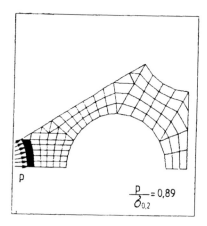
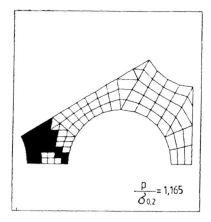
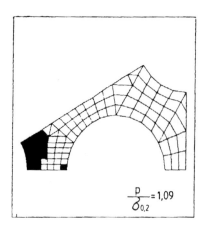
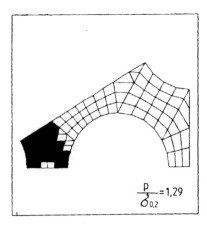

Fig. 5.27 Form of the plastic zone $t/D_i = 1.4$.

Methods of Fastening Tubes in Tubesheets and Headers

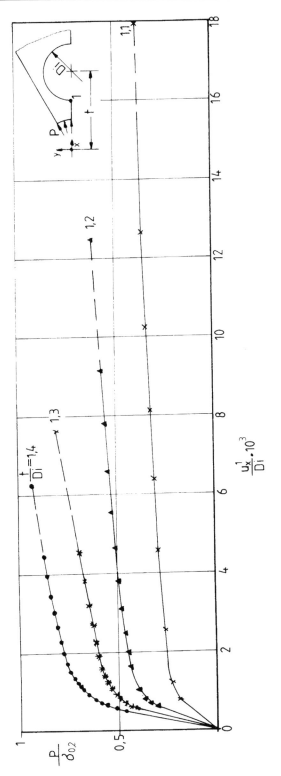

Fig. 5.28 Radial displacement of the adjacent borehole.

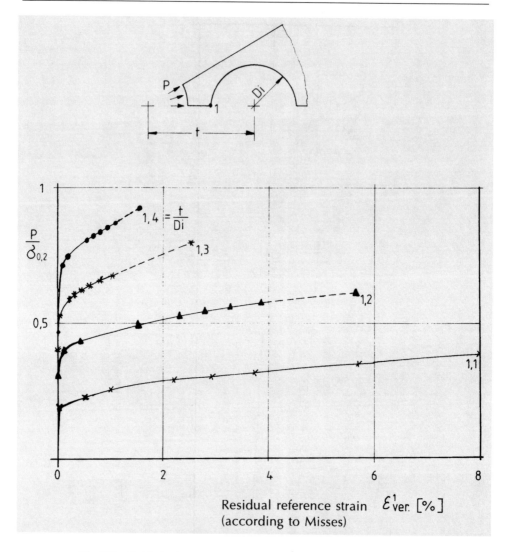

Fig. 5.29 Residual reference strain according to Misses of the adjacent borehole.

As can be seen from Fig. 5.31, the adhesion pressure at first creeps very quickly after warming to 900°C. After 15 hours it is only 100 bar. However, as it becomes more and more stable, the creep rate decreases continuously and goes asymptotically to a value of 50 bar.

This adhesion pressure is certainly not sufficient to seal the joint but it can adequately fulfill other functions which are required of tube/tubesheet joints in high temperature equipment, e.g., transfer of the tube load.

The discharge of the adhesion pressure during operation is no doubt very great in the case of constructions which operate at high temperatures and whose temperatures are in the creep range. This is also significant in the case of tube/tubesheet joints which do not creep but which are influenced by different rates of thermal expansion of the tube and the tubesheet as a result of transient pressure and temperature loads.

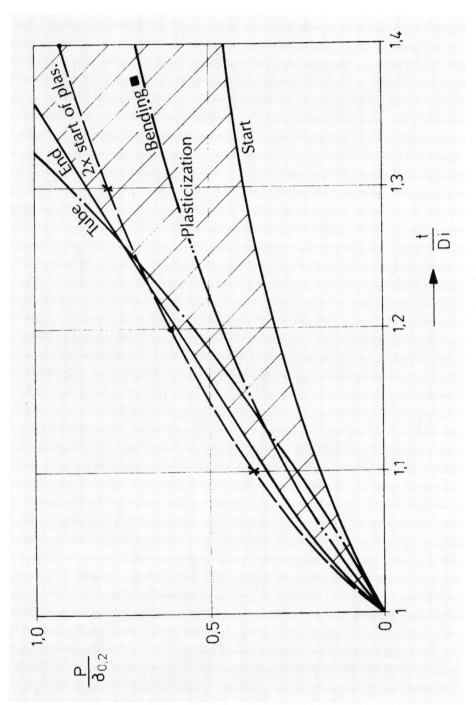

Fig. 5.30 Plasticization areas as a function of the tubesheet geometry.

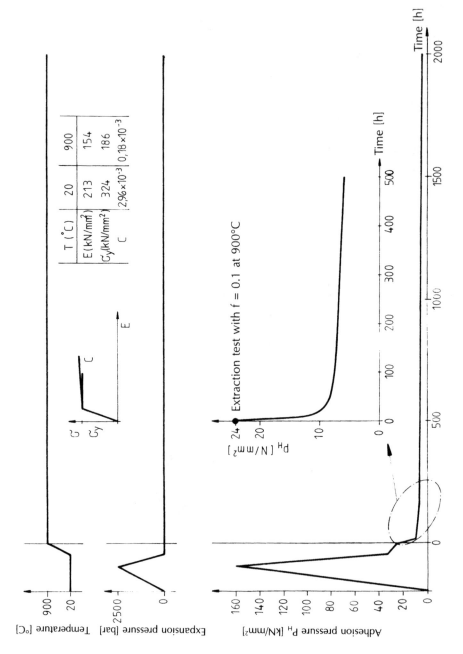

Fig. 5.31 Adhesion pressure as a function of time during a creep process.

Methods of Fastening Tubes in Tubesheets and Headers

The adhesion pressure built up at ambient temperature, increases or decreases depending on the load and material combination. The adhesion pressure is then a function of time. It is important to know whether the adhesion pressure in installed state is adequate to absorb the possible decrease in adhesion pressure during the individual load cases to ensure that adhesion is adequate over the entire service life of the component.

It is easy to take into account the time-variable internal pressure p and this can be done by inserting a single term in the equation to determine a fictitious clearances between the tube and the tubesheet.

It is more difficult with the temperature effect. It is necessary to determine the average wall temperatures of the tube T_{Rm} and of the tubesheet T_{Pm} as a function of time in order to calculate the individual radial temperature displacements. It is assumed that the heat transfer takes place only in a radial direction. This applies in particular to thicker tubesheets. The fictitious clearance is then calculated with

$$u_R = \frac{r_a}{r_i} \quad \text{and} \quad u_P = \frac{t}{r_a} - 1 :$$

$$s = \frac{2 p r_a}{E_R (u_R^2 - 1)} + \frac{\overline{p_H} r_a}{E_R} \frac{(1+\mu)}{(1 - 1/u_R^2)} \left(\frac{1}{u_R^2} + \frac{1-\mu}{1+\mu} \right)$$

$$+ \frac{\overline{p_H} r_a}{E_P} \frac{1+\mu}{u_P^2 - 1} \left(u_P^2 + \frac{1-\mu}{1+\mu} \right) + \alpha_R r_a \left(T_{Rm} - T_{R0} \right)$$

$$- \alpha_P r_a \left(T_{Pm} - T_{P0} \right) \tag{5.23}$$

In equation (5.23) $\overline{p_H}$ represents the initial adhesion pressure in the installed state. The fictitious clearance s can have either a positive or a negative value. With a positive clearance the tube and borehole are in contact, with a negative clearance there is a gap.

The average wall temperatures T_{Rm} and T_{Pm} can, for example be determined using the differential method [3]. The initial temperatures T_{R0} and T_{P0} are the tube and tubesheet temperatures at which the tube is fastened to the tubesheet. Fig. 5.32 shows the temperature pattern and the significance of the mean temperatures.

The thermal expansion coefficients of the tube and of the tubesheet material are designated by α_R and α_P.

The new time-related adhesion pressure p_H is then calculated as follows:

$$p_H = \frac{s}{\frac{r_a}{E_R} \frac{(1-\mu)}{(1-1/u_R^2)} \left(\frac{1}{u_R^2} + \frac{1-\mu}{1+\mu} \right) + \frac{r_a}{E_P} \frac{1+\mu}{u_P^2 - 1} \left(u_P^2 + \frac{1-\mu}{1+\mu} \right)} . \tag{5.24}$$

The internal pressure naturally increases the adhesion pressure p_H. The temperature then has a positive influence on the adhesion pressure if the mean tube wall temperature is higher than the mean tubesheet temperature. The thermal expansion coefficients α also exert a significant influence. The higher the thermal expansion coefficient of the tube

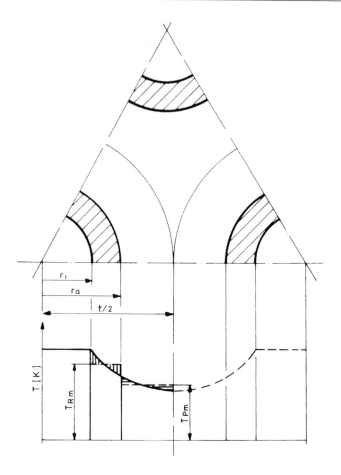

Fig. 5.32 Temperature pattern in the tube and the tubesheet.

material α_R is in comparison to the thermal expansion coefficient of the tubesheet, the quicker the adhesion pressure rises as the temperature of the medium in the tube increases.

The geometries of the tube and of the tubesheet influence the adhesion as well as the material properties.

Fig. 5.33 compares the adhesion pressure patterns of joints made of different material combinations at the same rate of temperature and pressure change. From this it is obvious that the influence of the thermal expansion coefficients of the tube and of the tubesheet are dominant during the start-up and shutdown processes.

With tubes made of St 35.8 and Incoloy 800 H the thermal expansion coefficient is greater than that of the tubesheet; with titanium it is the reverse. The influence of the intermittent temperature distribution in the tube and tubesheet is particularly noticeable in the variant with titanium. The relatively low internal pressure has only a negligible influence on the adhesion pressure pattern.

Finally, the special features of a possible combination of expansion/rolling-in should be mentioned. After hydraulic expansion the tube is perceptibly shorter and after rolling-

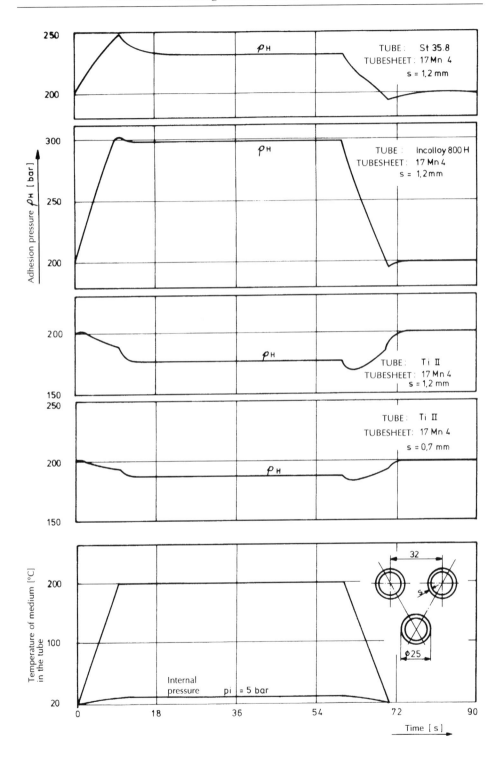

Fig. 5.33 Time-related adhesion pressure pattern for joints made of different material combinations.

in perceptibly longer. The reason for this is that during hydraulic expansion the tube yields, and there is virtually no reduction in wall thickness, whereas during the rolling process the tube material is rolled out and displaced.

This attribute can be used to great advantage in heat exchangers with straight tubes without an expansion joint in the shell in order to achieve a required prestressing of the tubes for the combined expansion/rolling-in process. In exceptional cases it is possible to fasten the tube stress-free in two rigid tubesheets.

Fig. 5.34 presents the diagram for the combined expansion/rolling-in process. The ratio of the clearance to the tube shortening as a result of hydraulic expansion, and of the tube extension to the rolling setting used in the rolling process, must be established in a test using the unit before work commences. The original tubes and rolling unit to be employed should be used for this to obtain transferable results. Fig. 5.34 shows, for example, that the suitable rolling setting for the mean clearance is determined from the two functions when fastening tubes stress-free. In this way, the two tube length changes cancel each other out.

The mean clearance is defined as the difference between the typical borehole diameter and the typical tube outer diameter. The typical diameters are the arithmetical mean values of a statistically representative number of measurements which can be easily determined by measuring the boreholes and the tubes.

The hydraulic expansion process is a relatively new method which has numerous advantages over the conventional means of fastening tubes in tubesheets by roller expansion. The expansion of the tube takes only seconds and irrespective of the fastening length there is no wear on the tools thus saves costs. Only one operator is required for the unit during the expansion process. Tube cleaning is not required.

In summary, the advantages of hydraulic expansion are:

- time and cost savings
- calculability
- uniformity of all joints
- minimal residual stresses in tube
- closing of gap between tube and tubesheet
- reliable monitoring
- possibility to control tube force

There is another type of expansion process, known as hydraulic expansion, in which, however, the hydraulics are used only to generate the required axial force. A patent application has been submitted for this process in the U.S. and Japan. Nothing has been published, however, on the practical applications.

The principle of the expansion process can be seen from Fig. 5.35. An axial force is generated using a hydraulic cylinder and transferred into a hollow cylinder made of rubber, silicon rubber or another elastomer. The axial shortening of the cylinder causes radial bulging of the flexible cylinder. This radial deformation naturally involves radial pressure on the tube wall. This pressure must be great enough to cause the plastic deformation of the tube and, in addition, similarly to the expansion process using a fluid, the tubesheet must be deformed to the extent that the tube remains firmly clasped after reverse elastic deformation.

The tube is not expanded uniformly over the entire expansion range. As a result, the joint is not uniform and it is not possible to predict its exact position, i.e., beginning

Methods of Fastening Tubes in Tubesheets and Headers

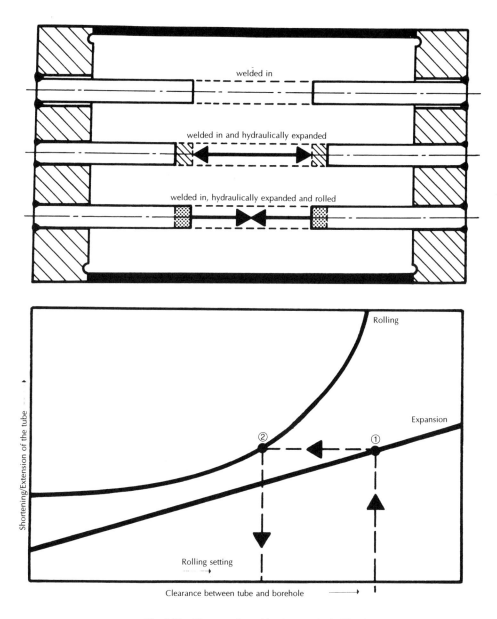

Fig. 5.34 Diagram of combined expansion/rolling-in.

and end. This method is not suitable for tubes with a small inside diameter or thick-walled tubes.

The fatigue and destruction of the flexible cylindrical core depends on the clearance and on the required axial force. It is time-consuming and costly to exchange the core. This process offers no advantages over the hydraulic expansion process using a fluid and has not become generally accepted.

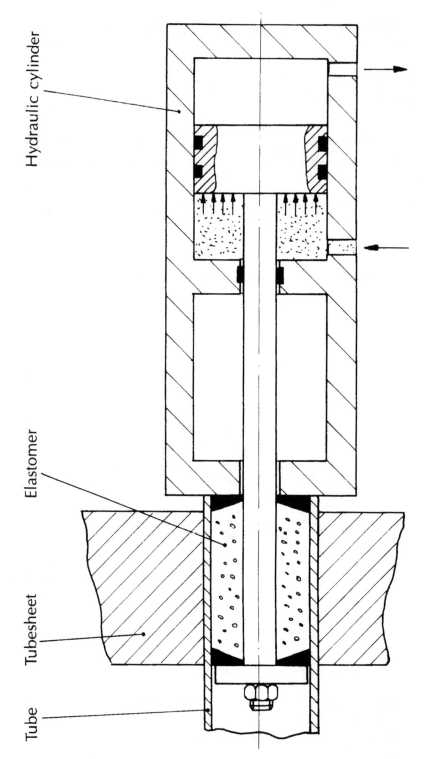

Fig. 5.35 Principle of the expansion process using an elastomer.

5.5 FASTENING OF TUBES BY EXPLOSION

This method of fastening tubes with the help of explosion was developed in the 1960s but to date it has been applied only to a small number of components. It is most frequently used to plug damaged tubes with difficult access. The term "fastening of tubes by explosion" encompasses in principle two different methods of tube fastening, namely, expansion where there is a clear separation between the tube material and the plate material, and explosion welding of the tubes, which inseparably joins the two materials. The two variants differ only in the velocity at which the two parts collide during the explosion.

The process of explosion welding used for tubes is based on the elementary process of the diagonal collision of surfaces at a sufficiently high velocity. The collision process can be illustrated simply with two flat plates (Fig. 5.36). The trajectory plate is held at a specified distance from the base plate by spacers. The air space corresponds approximately to the thickness of the trajectory plate on which an explosive layer is placed. The detonation velocity of the explosive substance is between 1,500 and 4,000 m/s.

The charge is ignited from one side so that an even detonation surface is obtained. A detonation shock wave then spreads which accelerates the trajectory plate locally. It collides with the plate diagonally at a velocity of several 100 m/s. A narrow, forward flowing material stream whose formation is regarded as an essential condition for the joint,

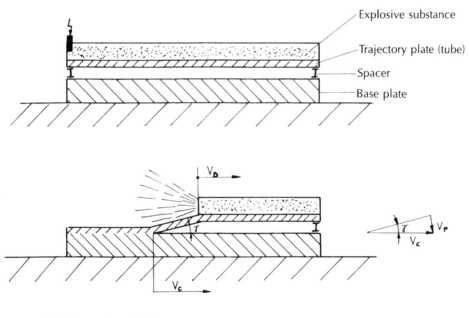

V_p - trajectory plate velocity
V_a - detonation velocity
V_c - collision velocity
γ - collision angle

Fig. 5.36 Principle of explosion welding.

forms in layers close to the surface as a result of this collision which spreads at the detonation velocity of the explosive substance.

The collision process can be described by three parameters: the trajectory plate velocity V_p, the collision velocity V_c and the collision angle γ. These three parameters are linked by the equation

$$v_p = 2 v_c \cdot \sin\frac{\gamma}{2} \qquad (5.25)$$

and must therefore be within certain characteristic limiting values for each combination of materials.

In practice, the optimum range is established through tests in which the properties and the mass grouping of the explosive substance, as well as the geometry and plate arrangement are varied.

A differentiation can be made between four characteristic types of explosive welds:

- flat joint without thermal structural change
- undulating joint without thermal structural change
- undulating joint with local semi-fusion
- flat joint with a continuous fusion covering.

The first two types are regarded as being favourable and are used wherever possible. The joint is formed by joining two surfaces at distances corresponding to the grid constants.

A surface-cleaning material stream forms on the collision line (Fig. 5.37) and is followed immediately by the high collision pressure. The collision pressure reaches values of approx. 130 Kbar and more. The shear strain produced by the formation of the material stream should be regarded as the main heat source. This shear strain is adiabatic and normally causes the semi-fusion of a covering, in most cases only 1 to 2 µm thick, along the joint line. If this impact energy is too large, then local semi-fusions occur or even a continuous fusion covering is created.

This joint may, however, be undesirable because a brittle intermediate covering or an intermediate coverings with initial cracks forms as a result of the rapid solidification of the fluid phase.

For a long time it was unclear whether or not penetration of the structure and structural change would occur at the material limits. With the help of structural analyses carried out with a transmission electron microscope, it could be shown that the interaction of the welded materials extended over a measurably wide area.

The structural changes in the weld section are illustrated in Fig. 5.37.

The strain-induced fusion is fundamental to the explosion fusion.

The advantage is that the material combinations which cannot be welded using other methods can be welded using this process. The disadvantage of this method is the possible material damage, such as kinking and impact caused by the high strain, in particular by the primary shock wave.

The increase in hardness at the surface and naturally also in the joint section is one of several unfavourable factors which accompany this process.

Fig. 5.38 shows one defective joint and one flaw-free undulating tube/tubesheet joint in which both parts were made of the material 10 Cr Mo 910.

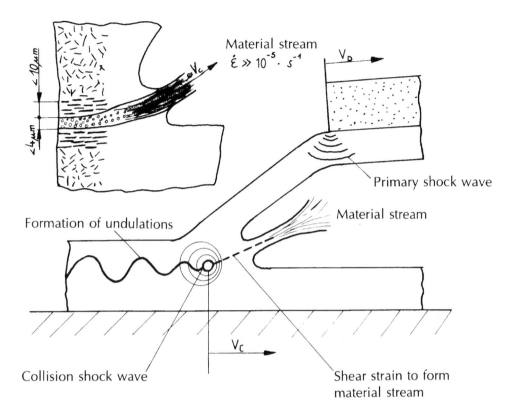

Fig. 5.37 Diagram of the collision process.

Up to now the explosion process has been dealt with in a general manner. It is used above all for the explosion-cladding of plates. When fastening tubes by explosion, certain constructional changes have to be carried out to enable the mechanism to work.

Some possible tubesheet types are shown in Fig. 5.39. The disadvantage of this construction is, however, that in certain cases larger pitches have to be selected. This depends on the ratio of the ligament width to the tube wall thickness.

The flow conditions are, naturally, significantly improved by the increase in the tube diameter. In the case of thick-walled tubes it is possible to create the required clearance for explosion welding by conically or cylindrically tapering the tube wall thickness without changing the borehole. This constructional solution should, of course, be always given preference because it does not weaken the ligament and, consequently, the explosive charge is lower. Unfortunately, it cannot be used in the case of thin-walled tubes. The positioning of the explosive charge must be very carefully considered so as to avoid buckling on the rear side of the tubesheet. In certain cases, a plug is inserted in the tube to prevent buckling. The plug exactly determines the expansion section and its mass prevents the spread of the detonation wave.

When plugging damaged tubes in cases in which the speed of the repair is important or the point to be repaired is difficult to reach, explosion plugs are used.

Fig. 5.40 illustrates two difference types of explosion plugs.

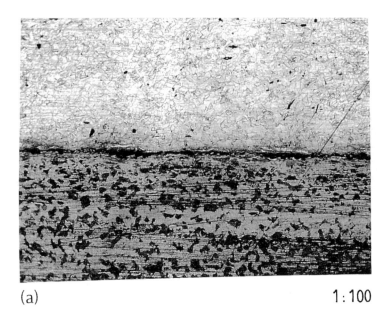

(a) 1:100

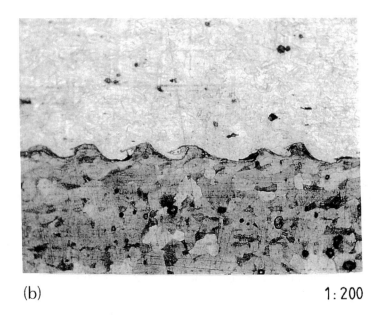

(b) 1:200

Fig. 5.38 Explosion weld joint, tube and tubesheet 10 Cr Mo 910 (1:200): (a) defective joint with interruptions; (b) undulating joint without flaw.

Methods of Fastening Tubes in Tubesheets and Headers

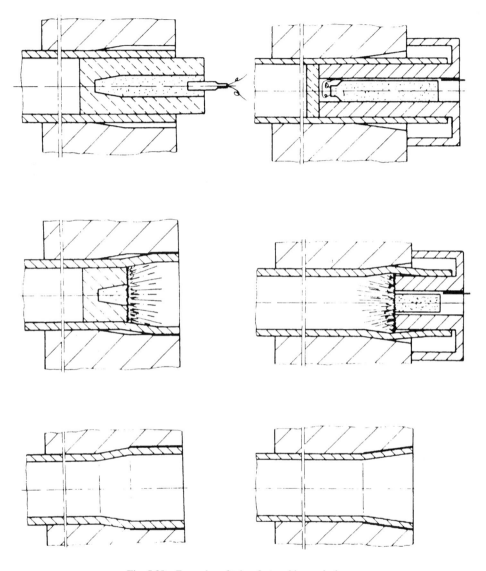

Fig. 5.39 Examples of tubes fastened by explosion.

Over the past few years the electrohydraulic tube fastening method has been offered in addition to the explosion method. This method is based on electrohydroimpulsive processes brought about by high-voltage discharge in the fluid. The high pressure pulses generated in the fluid in the tube cause local disturbances in the metal, similar to those caused in the explosion process, which then produce a dentiform joint. Pressure pulses and the corresponding high-speed plastic deformation reach velocities of between 30 and 400 m/s.

The booster circuit of an electrohydraulic pulse generator comprises a rectifier and a transformer to increase the voltage, and it enables the pulse to be released at the correct moment. The unit consists of the power supply to the electrode and a cartridge in which the electrode is positioned. Fig. 5.41 is a diagram of the unit.

A separate room, an operating room and a work room are needed to set up the unit, requiring considerable capital expenditure and resulting in a very inconvenient production flow.

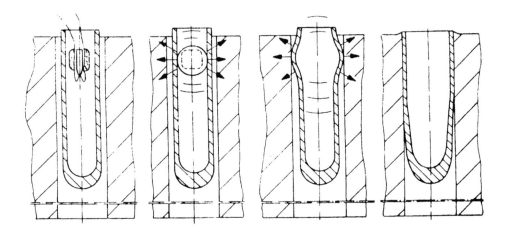

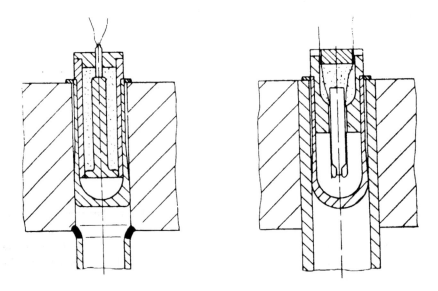

Fig. 5.40 Examples of explosive tube plugging.

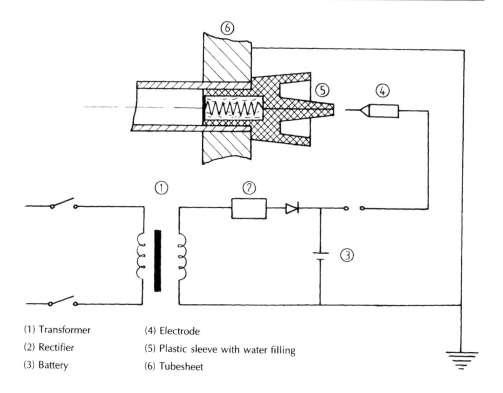

(1) Transformer
(2) Rectifier
(3) Battery
(4) Electrode
(5) Plastic sleeve with water filling
(6) Tubesheet

Fig. 5.41   Diagram of the electrohydraulic pulse generator.

BIBLIOGRAPHY

1. Krips, H. and Podhorsky, M., Hydraulisches Aufweiten - ein neues Verfahren zur Befestigung von Rohren, *VGB-KRAFTWERKSTECHNIK*, 59 (1976), Heft 7.
2. Podhorsky, M. and Krips, H., Hydraulisches Aufweiten von Rohren, *VGB-KRAFTWERKSTECHNIK*, 59 (1979), Heft 1.
3. Podhorsky, M., Ein Beitrag zur Ermittlung der instationären Wärmespannungen in Rohrplatten, *VGB-KRAFTWERKSTECHNIK*, 60 (1980), Heft 8.
4. Podhorsky, M., Design of Modern Heat Exchangers Using Hydraulic Tube Expansion, 6-ICPVT, Beijing, September 1988.
5. Hardwick, R., Methods for Fabricating and Plugging of Tube-to-Tubesheet Joints by Explosion Welding, *Welding Journal*, April 1975.
6. Scott, D. A., Wolgemuth, G. A., and Aikin, J. A., Hydraulic Expanded Tube-to-Tubesheet Joint, *Journal of Pressure Vessel Technology*, Vol. 106, February 1984.
7. Gaffoglio, C. J. and Thiele, E. W., Tube-to-Tube-Sheet Joint Strengths, ASME-Paper 81-JPGC-Pur-7.
8. Fino, A. F. and Dumas, W. A., Hydraulic Expansion System Produces Leak-Free Tube-to-Tubesheet Joints, Power, October 1979.
9. Yokell, S., Hydroexpanding: the Current State of the Art, ASME 82-JPGC-PWR-!, *ASME*, New York, 1982.
10. Aldred, D. L., Hydraulic Expansion of Tube to Tubesheet Joints, Proc. 1st U.K. National Heat Transfer Conf., Leeds, U.K., 1984.

11. Nadai, H., Theory of the Expanding of Boiler and Condenser Tube Joints Through Rolling, ASME Spring meeting, Davenport, Iowa, April 1943, ASME, New York.
12. Goodier, J. N. and Schoessow, G. J., The Holding Power and Hydraulic Tightness of Expanded Tube Joints, *Trans ASME,* 489, July 1943, ASME, New York.
13. Sachs, G., Notes on the Tightness of Expanded Tube Joints, *J. Appl, Mech.,* A-285, Dec. 1947, ASME, New York.
14. Jackson, P. W., Heat Exchanger Repair Techniques Involving the Use of Explosives, Repair & Reclamation Conf., London, Sept. 1984.
15. Hardwick, R., Tube/Tubeplate Joining, Plugging and Other Explosive Techniques Which Can Be Used in the Construction and Repair of Heat Exchangers, 8th Int. Conf. on High Energy Rate Fabrication, San Antonio, June 17–21, 1984.

APPENDIX

EXAMPLES FOR CALCULATION

Chapter: 2.3 Flat Heads

The significance of the individual letters can be seen from Fig. 2.1.

$$R = 300 \text{ mm} \qquad E = 2 \cdot 10^5 \text{ N/mm}^2 \qquad T_m = T_0$$
$$t_s = 20 \text{ mm} \qquad \mu = 0.3$$
$$t = 60 \text{ mm} \qquad p_i = 150 \text{ bar}$$

$$w = \frac{2F_3}{3E(t/R)} \cdot Q + \frac{F_3}{E \cdot R(t/R)^2} \cdot M - p_i \frac{t}{2} \frac{F_1}{E(t/R)^3} \tag{2.1}$$

$$\Theta = \frac{F_3}{ER(t/R)^3} \cdot Q + \frac{2F_3}{E \cdot R^2(t/R)^3} \cdot M - p_i \frac{F_1}{E(t/R)^3} \tag{2.1}$$

$$\frac{t}{R} = \frac{60}{300} = 0.2$$

$$R_m = R - \frac{t_s}{s} = 300 - \frac{20}{2} = 290 \text{ mm}$$

$$f = \frac{t_s}{R} = -\frac{20}{300} = 0.066 \text{ mm}$$

$$F_1 = \frac{3(1 - \mu)(2 - f^2)(1 - f^2) \cdot [8 - f(4 - f)(1 - \mu)]}{16 \cdot (2 - f)} = 1.054 \tag{2.2}$$

$$F_2 = \frac{3}{8}(1 - f^2) \cdot \left[(1 - \mu)(2 - f^2) + 4(1 + \mu)\left(1 + 2 \cdot \ln\frac{2 - f}{2 - 2f}\right)\right] = 2.17 \tag{2.2}$$

$$F_3 = \frac{3}{8}(1 - \mu)(2 - f) \cdot [8 - f(4 - f)(1 - \mu)] = 3.969 \tag{2.2}$$

$$F_4 = \frac{1}{8}[8 - f(4 - f)(1 - \mu)] = 0.977 \tag{2.2}$$

$$w = \frac{2 \cdot 3.969}{3 \cdot 2 \cdot 10^5 \cdot 0.2} \cdot Q + \frac{3.969}{2 \cdot 10^5 \cdot 300 \cdot 0.2^2} \cdot M - 15 \cdot \frac{60}{2} \frac{1.054}{2 \cdot 10^5 \cdot 0.2^3}$$

$$w = 6.615 \cdot 10^{-5} \cdot Q + 0.1653 \cdot 10^{-5} \cdot M - 0.296$$

$$\Theta = \frac{3.969}{2 \cdot 10^5 \cdot 300 \cdot 0.2^2} \cdot Q + \frac{2 \cdot 3.969}{2 \cdot 10^5 \cdot 300^2 \cdot 0.2^3} \cdot M - 15 \cdot \frac{1.054}{2 \cdot 10^5 \cdot 0.2^3}$$

$$\Theta = 0.165 \cdot 10^{-5} \cdot Q + 0.0055 \cdot 10^{-5} \cdot M - 0.00988$$

For $x = \dfrac{t}{2}$ and $r = 0$:

$$\sigma_r = \frac{F_4}{t}\left(1 - \frac{6x}{t}\right) \cdot Q - \frac{12 \cdot F_4 \cdot x}{t^3} \cdot M + \frac{x \cdot p_i}{t(t/R)^2} \cdot F_2 \qquad (2.3)$$

$$\sigma_r = \frac{0.977}{60}(1 - 3) \cdot Q - \frac{6 \cdot 0.977}{60^2} \cdot M + \frac{15}{2 \cdot 0.2^2} \cdot 2.17$$

$$\sigma_r = -0.0325 \cdot Q - 0.0016 \cdot M + 406.8$$

$$\sigma_t = \frac{F_4}{t}\left(1 - \frac{6x}{t}\right) \cdot Q - \frac{12 \cdot F_4 \cdot x}{t^3} \cdot M + \frac{x \cdot p_i}{t(t/R)^2} \cdot F_2 \qquad (2.3)$$

$$\sigma_t = -0.0325 \cdot Q - 0.0016 \cdot M + 406.8$$

$$\sigma_a = \left(x - \frac{t}{2}\right) \cdot \frac{p_i}{t} \qquad (2.3)$$

$$\sigma_a = 0$$

Chapter: 2.5 Spherical Shells

The significance of the individual letters can be seen from Fig. 2.1.

$R = 1000$ mm $\qquad E = 2 \cdot 10^5$ N/mm² $\qquad p_i = 50$ bar
$t = 20$ mm $\qquad \mu = 0.3$ $\qquad T_m = T_0$

$$w = \frac{2R\lambda}{Et} \cdot Q + \frac{2\lambda^2}{Et} \cdot M + p_i \frac{2R^3(1 - 2\mu) + (R + t/2)^3(1 + \mu)}{2ER^2(u^3 - 1)} \qquad (2.13)$$

$$\Theta = \frac{2\lambda^2}{Et} \cdot Q + \frac{4\lambda^3}{ERt} \cdot M \tag{2.13}$$

$$\lambda = \beta \cdot R = 0.009 \cdot 1000 = 9 \tag{2.14}$$

$$\beta = [3(1-\mu^2)]^{1/4} \frac{1}{\sqrt{Rt}} = (3 \cdot 0.91)^{1/4} \frac{1}{\sqrt{1000 \cdot 20}} = 0.009 \tag{2.14}$$

$$u = \frac{R + t/2}{R - t/2} = \frac{1000 + 10}{1000 - 10} = 1.02 \tag{2.14}$$

$$w = \frac{2 \cdot 1000 \cdot 9}{2 \cdot 10^5 \cdot 20} \cdot Q + \frac{2 \cdot 9^2}{2 \cdot 10^5 \cdot 20} \cdot M + 5\frac{2 \cdot 1000^3 \cdot 0.4 + 1010^3 \cdot 1.3}{2 \cdot 2 \cdot 10^5 \cdot 1000^2(1.02^3 - 1)}$$

$$= 4.5 \cdot 10^{-3} \cdot Q + 4.05 \cdot 10^{-5} \cdot M + 0.0873$$

$$\Theta = \frac{2 \cdot 9^2}{2 \cdot 10^5 \cdot 20} \cdot Q + \frac{2 \cdot 9^3}{2 \cdot 10^5 \cdot 1000 \cdot 20} \cdot M = 4.05 \cdot 10^{-5} \cdot Q$$

$$+ 7.29 \cdot 10^{-7} \cdot M$$

Chapter: 2.6 Cylindrical Shells

The significance of the individual letters can be seen from Fig. 2.3. For the long cylindrical shell is

$$B_{11} = B_{12} = B_{22} = 1 \quad R = 1000 \text{ mm} \quad E = 2 \cdot 10^5 \text{ N/mm}^2$$
$$G_{11} = G_{12} = G_{22} = 1 \quad t = 20 \text{ mm} \quad \mu = 0.3$$
$$p_i = 20 \text{ bar} \quad T_m = T_0$$

$$w_L = -\frac{1}{2\beta^3 D} \cdot Q_L + \frac{1}{2\beta^2 D} \cdot M_L + p_i \frac{(1 - \mu/2)R(R - t/2)}{Et} \tag{2.18}$$

$$\Theta_L = -\frac{1}{2\beta^2 D} \cdot Q_L + \frac{1}{2\beta D} \cdot M_L \tag{2.18}$$

$$D = \frac{Et^3}{12(1-\mu^2)} = \frac{2 \cdot 10^5 \cdot 20^3}{12 \cdot 0.91} = 1.465 \cdot 10^8 \text{ [Nmm]} \tag{2.24}$$

$$\beta = [3(1-\mu)]^{1/4} \frac{1}{\sqrt{Rt}} = (3 \cdot 0.7)^{1/4} \cdot \frac{1}{\sqrt{1000 \cdot 20}} = 0.0085 \tag{2.24}$$

$$w_L = \frac{1}{2 \cdot 0.0085^3 \cdot 1.465 \cdot 10^8} Q_L + \frac{1}{2 \cdot 0.0085^2 \cdot 1.465 \cdot 10^8} \cdot M_L$$

$$+ 2 \cdot \frac{0.85 \cdot 1000 \cdot 990}{2 \cdot 10^5 \cdot 20}$$

$$w_L = -1.11 \cdot 10^{-2} Q_L + 9.4 \cdot 10^{-5} \cdot M_L + 0.42$$

$$\Theta_L = -\frac{1}{2 \cdot 0.0085^2 \cdot 1.465 \cdot 10^8} Q_L + \frac{1}{2 \cdot 0.0085 \cdot 1.465 \cdot 10^8} \cdot M_L$$

$$\Theta_L = -9.4 \cdot 10^{-5} Q_L + 2.36 \cdot 10^{-6} \cdot M_L$$

Chapter: 2.9 Rings

The significance of the individual letters can be seen in Fig. 2.6.

$$\begin{array}{lll} r_a = 1200 \text{ mm} & S_L = S_R = 20 \text{ mm} & p_i = 20 \text{ bar} \\ r_i = 20 \text{ mm} & E = 2 \cdot 10^5 \text{ N/mm}^2 & T_m = T_0 \\ h = 300 \text{ mm} & \mu = 0.3 & r_L = r_R = r_i \end{array}$$

$$w_L = (A + B) \cdot Q_L + C \cdot M_L + (-A + B) \cdot Q_R - C \cdot M_R$$

$$+ p_i \left[\frac{h \cdot r_i^2 (a_L - a_r)}{4ED_R} + \frac{r_i (r_a + r_i)}{2E(r_a - r_i)} \right] \tag{2.40}$$

$$\Theta_L = C \cdot Q_L + \frac{2C}{h} \cdot M_L - C \cdot Q_R - \frac{2C}{h} \cdot M_R + p_i \frac{r_i^2 (a_L - a_R)}{2ED_R} \tag{2.40}$$

$$w_R = (-A + B) \cdot Q_L - C \cdot M_L + (A + B) \cdot Q_R + C \cdot M_R$$

$$+ p_i \left[\frac{-h \cdot r_i^2 (a_L - a_R)}{4ED_R} + \frac{r_i (r_a - r_i)}{2E(r_a - r_i)} \right] \tag{2.40}$$

$$\Theta_R = \Theta_L \tag{2.40}$$

$$D_R = \frac{h^3}{12} \cdot \ln \frac{r_a}{r_i} = \frac{300^3}{12} \cdot \ln \frac{1200}{1000} = 4.1 \cdot 10^5 \, [\text{mm}^3] \tag{2.41}$$

Appendix

$$a_L = \frac{(r_a + r_i)}{2} - \left(r_L + \frac{s_L}{2}\right) = \frac{(1200 + 1000)}{2} - \left(1000 + \frac{20}{2}\right) = 90 \text{ [mm]} \quad (2.41)$$

$$a_R = a_L \quad (2.41)$$

$$A = \frac{h^2}{8 \cdot \pi \cdot E \cdot D_R} = \frac{300^2}{8 \cdot \pi \cdot 2 \cdot 10^5 \cdot 4.1 \cdot 10^5} = 4.36 \cdot 10^{-8} \left[\frac{mm}{N}\right] \quad (2.41)$$

$$B = \frac{r_a + r_i}{4 \cdot \pi \cdot h \cdot E \cdot (r_a - r_i)} = \frac{1200 + 1000}{4 \cdot \pi \cdot 300 \cdot 2 \cdot 10^5 (1200 - 1000)}$$

$$= 1.46 \cdot 10^{-8} \left[\frac{mm}{N}\right] \quad (2.41)$$

$$C = \frac{h}{4 \cdot \pi \cdot E \cdot D_R} = \frac{300}{4 \cdot \pi \cdot 2 \cdot 10^5 \cdot 4.1 \cdot 10^5} = 2.912 \cdot 10^{-10} \left[\frac{mm}{N}\right] \quad (2.41)$$

$$w_L = (4.36 \cdot 10^{-8} + 1.46 \cdot 10^{-8})Q_L + 2.912 \cdot 10^{-10} \cdot M_L + (-4.36 \cdot 10^{-8}$$
$$+ 1.46 \cdot 10^{-8})Q_R - 2.912 \cdot 10^{-10} \cdot M_R + 2 \cdot \frac{1000 \cdot (1200 + 1000)}{2 \cdot 2 \cdot 10^5 (1200 - 1000)}$$

$$w_L = 5.82 \cdot 10^{-8} \cdot Q_L + 2.912 \cdot 10^{-10} \cdot M_L - 2.9 \cdot 10^{-8} \cdot Q_R - 2.912$$
$$\cdot 10^{-10} \cdot M_R + 0.055$$

$$\Theta_L = 2.912 \cdot 10^{-10} \cdot Q_L + \frac{2.912 \cdot 10^{-10}}{300} \cdot M_L - 2.912 \cdot 10^{-10} \cdot Q_R - \frac{2 \cdot 2.912 \cdot 10^{-10}}{300} \cdot M_R$$

$$\Theta_L = 2.912 \cdot 10^{-10} \cdot Q_L + 1.94 \cdot 10^{-12} \cdot M_L - 2.912 \cdot 10^{-10} \cdot Q_R - 1.94 \cdot 10^{-12} \cdot M_R$$

$$w_R = -2.9 \cdot 10^{-8} \cdot Q_L - 2.912 \cdot 10^{-10} \cdot M_L + 5.82 \cdot 10^{-8} \cdot Q_R$$
$$+ 2.912 \cdot 10^{-10} \cdot M_R + 0.055$$

$$\Theta_R = \Theta_L$$

Chapter: 4.5. Dimensioning of Flanges Using the Deformation Calculation

The significance of the individual letters can be seen in Fig. 4.14.

Appendix

$$d_i = 1000 \text{ mm} \qquad S_R = 20 \text{ mm} \qquad E = 2 \cdot 10^5 \text{ N/mm}^2$$
$$d_D = 1040 \text{ mm} \qquad h_F = 80 \text{ mm} \qquad \mu = 0.3$$
$$d_t = 1100 \text{ mm} \qquad F_S = F_D \qquad d_L = 22 \text{ mm}$$
$$d_F = 1160 \text{ mm} \qquad F_R = F_F = 0 \qquad t = 66 \text{ mm}$$

$$a_{11} \cdot T_1 + a_{12} \cdot M_1 = b_1 \tag{4.35}$$

$$a_{21} \cdot T_1 + a_{22} \cdot M_1 = b_2 \tag{4.35}$$

$$D_F = \frac{h_F^3}{12} \cdot \ln\frac{d_F}{d_i} = \frac{80^3}{12} \cdot \ln\frac{1160}{1000} = 6.33 \cdot 10^3 \left[\text{mm}^3\right] \tag{4.14}$$

$$\beta_Z = \frac{\sqrt[4]{3(1 - \mu^2)}}{\sqrt{\frac{(d_i + s_R)}{2} \cdot s_R}} = \frac{\sqrt[4]{3 \cdot 0.91}}{\sqrt{\frac{1020}{2} \cdot 20}} = 0.0127 \left[\frac{1}{\text{mm}}\right] \tag{4.32}$$

$$D_Z = \frac{s_R^3 \cdot E}{12(1 - \mu)} = \frac{20^3 \cdot 2 \cdot 10^5}{12 \cdot 0.7} = 1.9 \cdot 10^8 \text{ [Nmm]} \tag{4.32}$$

$$a_{11} = \frac{ED_F}{\beta_Z^2 \cdot D_Z(d_i + s_R)} - \frac{h_F}{2} = \frac{2 \cdot 10^5 \cdot 6.33 \cdot 10^3}{0.0127^2 \cdot 1.9 \cdot 10^8 \cdot (1000 + 20)} - \frac{80}{2} = 0.5 \tag{4.36}$$

$$a_{12} = -\frac{2ED_F}{\beta_Z \cdot D_Z(d_i + s_R)} - 1 = -\frac{2 \cdot 10^5 \cdot 6.33 \cdot 10^3}{0.0127^2 \cdot 1.9 \cdot 10^8 \cdot (1000 + 20)} - 1 = -1.514 \tag{4.36}$$

$$a_{21} = -\frac{h_F}{2} - \frac{(d_i + s_R)D_F}{h_F \cdot A_F} - \frac{2ED_F}{h_F(d_i + s_R) \cdot \beta^3 z \cdot D_Z} \tag{4.36}$$

$$A_F = A_1 \frac{1}{1 + e/t(A_1/A_2 - 1)} \tag{4.15}$$

$$A_1 = \frac{(d_F - d_i)h_F}{2} = \frac{(1160 - 1000) \cdot 80}{2} = 6400 \left[\text{mm}^2\right] \tag{4.15}$$

$$A_2 = e \cdot h_F = 22 \cdot 80 = 1760 \left[\text{mm}^2\right] \tag{4.15}$$

$$A_F = 6400 \frac{1}{1 + \frac{22}{66}\left(\frac{6400}{1760} - 1\right)} = 3406 \left[\text{mm}^2\right]$$

$$a_{21} = -\frac{80}{2} - \frac{(1000 + 20) \cdot 6.33 \cdot 10^3}{80 \cdot 3406} - \frac{2 \cdot 2 \cdot 10^5 \cdot 6.33 \cdot 10^3}{80(1000 + 20) \cdot 0.0127^3 \cdot 1.9 \cdot 10^8}$$

$$= -40 - 23.69 - 79.72 = -143.41$$

$$a_{22} = -1 + \frac{2 \cdot E \cdot D_F}{(di + s_R)\beta_z^2 h_F D_Z} = -1 + \frac{2 \cdot 2 \cdot 10^5 \cdot 6.33 \cdot 10^3}{(1000 + 20) \cdot 0.0127^2 \cdot 1.9 \cdot 10^8 \cdot 80}$$

$$= -1 + 1.0125 = 0.0125$$

$$b_1 = F_s \frac{(d_t - d_0)}{2} + F_D \frac{(d_0 - d_D)}{2} \tag{4.36}$$

$$d_0 = 2 \cdot \frac{\frac{d_F}{2} - \frac{d_i}{2} - \frac{d_L^2}{t}}{\ln\frac{d_F}{d_i} - \frac{d_L}{t} \cdot \ln\frac{d_t + d_L}{d_t - d_L}} = 2 \cdot \frac{580 - 500 - \frac{22^2}{66}}{\ln\frac{1160}{1000} - \frac{22}{66} \cdot \ln\frac{1100 + 22}{1100 - 22}} = 1075.8 \,[\text{mm}]$$

$$b_1 = F_s \cdot \left(\frac{1100 - 1075.8}{2} + \frac{1075.8 - 1040}{2}\right) = 30 \cdot F_s$$

$$b_2 = b_1 \tag{4.36}$$

$$M_1 = -19.88 \cdot F_s$$

$$T_1 = 0$$

ADDITIONAL BIBLIOGRAPHY

ANSI - American National Standards Institute

ANSI B 78.1 - 1982   Tubular Heat Exchangers in Chemical Process Service

API - American Petroleum Institute

API Publication 543   Heat Recovery Steam Generators, 1st Edition 1995
API Standard 660   Shell-and-Tube Heat Exchangers for General Refinery Services, 5th Edition, 1993
API Standard 661   Air-Cooled Heat Exchangers for General Refinery Services, 3rd Edition, 1992
API Standard 662   Plate Heat Exchangers for General Refinery Services, 1st Edition, 1995

API Guide for Inspection of Refinery Equipment, Chapter VII, Heat Exchangers, Condensers and Cooler Boxes

ASHRAE - American Society of Heating, Refrigerating and Air-Conditioning Engineers

Heating, Ventilating and Air-Conditoning. Systems and Equipment. Handbook, 1996

ASME - American Society of Mechanical Engineers

ASME Section I   Power Boilers - 1995 Edition
ASME Section VIII, Division 1   Rules for the Construction of Pressure Vessels - 1995 Edition
ASME Section VIII, Division 2   Rules for the Construction of Pressure Vessels - Alternative Rules - 1995 Edition

ASME - Fluids Engineering Division (FED)

1988 International Symposium on Flow-Induced Vibration and Noise

Vol. 1 Flow-Induced Vibration in Cylindrical Structures: Solitary Cylinders and Arrays in Cross-Flow, Paidoussis, M.P.; Griffin, O.M.; Dalton, C.
Vol. 2 Flow-Induced Vibration of Cylinder Arrays in Cross-Flow, Paidoussis, M.P.; Stininger, D.A.; Wambsganss, M.W.
Vol. 3 Flow-Induced Vibration and Noise in Cylinder Arrays, Paidoussis, M.P.; Chen, S.S.; Bernstein, M.D.
Vol. 4 Flow-Induced Vibrations due to Internal and Annular Flows and Special Topics in Fluidelasticity, Paidoussis, M.P.; Au-Yang, M.K.; Chen, S.S.

Vol. 5 Flow-Induced Vibration in Heat-Transfer Equipment, Paidoussis, M.P.; Chenoweth, J.M.; Chen, S.S.; Stenner, J.R.; Bryan, W.J.

Vol. 6 Acoustic Phenomena and Interaction in Shear Flows over Compliant and Vibrating Surfaces, Keith, W.L.; Uram, E.M.; Kalinowski, A.J.

Vol. 7 Nonlinear Interaction Effects and Chaotic Motions, Reischman, M.M.; Paidoussis, M.P.; Hansen, R.J.

ASME Pressure Vessel and Piping Division (PVP)

ASME PVP - Vol. 104	Flow-Induced Vibrations - 1986, Chen, S.S.; Simonis, J.C.; Shin, Y.S.
ASME PVP - Vol. 122	Flow-Induced Vibrations - 1987, Au-Yang. M.K.; Chen, S.S.
ASME PVP - Vol. 154	Flow-Induced Vibration - 1989, Au-Yang, M.K.; Chen, S.S.; Kaneko, S.; Chilukuri, R.
ASME PVP - Vol. 189	Flow-Induced Vibration - 1990, Chen, S.S.; Kujita, K.; Au-Yang, M.
ASME PVP - Vol. 194	Analysis of Pressure Vessel and Heat Exchanger Components - 1990, Short II, W.E.; Brooks, G.N.
ASME PVP - Vol. 206	Flow-Induced Vibration and Wear - 1991, Au-Yang, M.K.; Hara, F.
ASME PVP - Vol. 258	Flow-Induced Vibration and Fluid-Structure Interaction - 1993, Au-Yang, M.K.; Ma, D.C. et al.

ASME Heat Transfer Division (HTD)

ASME HTD-Vol. 9	Flow-Induced Heat Exchanger Tube Vibration - 1980, Chenoweth, J.M.; Stenner, J.R.
ASME HTD-Vol. 10	Compact Heat Exchangers, History, Technological Advancements and Mechanical Design Problems - 1980, Shah, R.K.; McDonald, C.F.; Howard, C.P.
ASME HTD-Vol. 14	Scaling in Two-Phase Flows - 1980, Saha, P.; Farukhi, N.M.
ASME HTD-Vol. 21	Regenerative and Recuperative Heat Exchangers - 1981, Shah, R.K.; Metzger, D.E.
ASME HTD-Vol. 34	Basic Aspects of Two-Phase Flow and Heat Transfer - 1984, Dhir, V.K.; Schrock, V.E.
ASME HTD-Vol. 35	Fouling in Heat Exchange Equipment - 1984, Suitor, J.W.; Pritchard, A.M.
ASME HTD-Vol. 36	A Reappraisal of Shellside Flow in Heat Exchangers - 1984, Marner, W.J.; Chenoweth, J.M.
ASME HTD-Vol. 43	Advances in Enhanced Heat Transfer - 1985, Shenkman, S.M.; O'Brien, J.E.; Habib, I.S.; Kohler, J.A.
ASME HTD-Vol. 44	Two-Phase Heat Exchanger Symposium - 1985, Pearson, J.T.; Kitto Jr., J.B.
ASME HTD-Vol. 49	Radiation Heat Transfer - 1985, Armaly, B.F.; Emery, A.F.
ASME HTD-Vol. 59	Heat Transfer in Waste Heat Recovery and Heat Rejection Systems - 1986, Chiou, J.P.; Sengupta, S.

Additional Bibliography

ASME HTD-Vol. 64 (PVP-Vol. 118)	Thermal/Mechanical Heat Exchanger Design - Karl Gardner Memorial Session - 1986, Singh, K.P.; Shenkmann, S.M.
ASME HTD - Vol. 66	Advances in Heat Exchanger Design - 1986, Shah, R.K.; Parson, J.T.
ASME HTD - Vol. 68	Advances in Enhanced Heat Transfer - 1987, Jensen, M.K.; Carey, V.P.
ASME HTD - Vol. 75	Maldistribution of Flow and its Effect on Heat Exchanger Performance - 1987
ASME HTD - Vol. 77	Nonequilibrium Transport Phenomena - 1987, Bankoff, S.G.; Chen, J.C.; El-Genk, M.S.
ASME HTD - Vol. 85	Boiling and Condensation in Heat Transfer Equipment - 1987, Ragi, E.G.; Rudy, T.M.; Rabas, T.J.; Robertson, J.M.
ASME HTD - Vol. 86	Effects of Fouling and Corrosion on Heat Transfer in Heat Rejection Systems - 1987, Mussalli, Y.
ASME HTD - Vol. 102	Thermal Hydraulics of Nuclear Steam Generators/Heat Exchangers - 1988, Hassan, Y.A.
ASME HTD - Vol. 136	Fundamentals of Phase Change: Boiling and Condensation - 1990, Witte, L.C.; Avedisian, C.T.
ASME HTD - Vol. 138	Heat Transfer in Turbulent Flow - 1990, Amano, R.S.; Crawford, M.E.; Anand, N.K.
ASME HTD - Vol. 159	Phase Change Heat Transfer - 1991, Hensel, E.; Dhir, V.K.; Greif, R.; Fillo, J.
ASME HTD - Vol. 169	Advances in Heat Transfer Augmentation and Mixed Convection - 1991, Ebadian, M.A.; Pepper, D.W.; Diller, T.K.
ASME HTD - Vol. 210	Fundamentals of Forced Convection Heat Transfer - 1992, Ebadian, M.A.; Oosthuizen, P.H.
ASME HTD - Vol. 213	Fundamentals of Mixed Convection - 1992, Chen, T.S.; Chu, T.Y.
ASME HTD - Vol. 217	Fundamentals of Subcooled Boiling - 1992, Boyd Sr., R.D.; Kandiikar, S.G.
ASME HTD - Vol. 220	Two-Phase Flow in Energy Exchange Systems - 1992, Sohal, M.S.; Raba, T.J.
ASME HTD - Vol. 221	Heat Pipes and Thermosysphons - 1992, Chang, W.S.; Gerner, F.M.; Ravigururajan, T.S.
ASME HTD - Vol. 222	Benchmark Problems for Heat Transfer Codes - 1992, Blackwell, B.; Pepper, D.W.
ASME HTD - Vol 285	Fundamentals of Heat Transfer in Forced Convection - 1994, Schmidt, F.W.; Moffat, R.J.
ASME HTD - Vol 287	Advances in Enhanced Heat Transfer - 1994, Rabas,T.J.; Bogart, J.E.
ASME HTD - Vol 314	1995 National Heat Transfer Conference, Vol. 12 - 1995, Sernas, V.; Boyd, R.D.; Jensen, M.K.

ASME - Nuclear Engineering Division (NE)

ASME NE - Vol. 1	Surface Condenser Design, Installation and Operating Experience - 1978

ASME NE - Vol. 5	Thermal Hydraulics of Advanced Heat Exchangers - 1990, Hassan, Y.A.; Cho, S.M.
ASME NE - Vol. 8	Steam Generator Sludge Disposition in Recirculating and Once-Through Steam Generator Upper Tube Bundle and Support Plates, Baker, R.L.; Harvey, E.A.
ASME NE - Vol. 15	Thermal Hydraulics of Advanced Steam Generators and Heat Exchangers - 1994, Hassan, Y.A.; Cassell, D.S.; Okamoto, K; Cho, M.

ASME PWR

ASME PWR Vol. 12	Performance Monitoring and Replacement of Heat Exchanger Components and Materials - 1990, Maurer, J.R.
ASME PWR Vol. 14	Practical Aspects and Performance of Heat Exchanger Components and Materials - 1991, Maurer, J.R.
ASME PWR Vol. 19	Practical Aspects and Performance of Heat Exchanger Components and Materials, Maurer, J.R.

ASME - 1992 International Symposium on Flow-Induced Vibration and Noise

ASME HTD - Vol. 230 (NE-Vol. 9)	FSI/FIV in Cylinder Arrays in Cross-Flow - Vol. 1, Paidoussis, M.P.; Bryan, W.J.; Stenner, J.R.; Steininger, D.A.
ASME PVP - Vol. 242	Cross-Flow Induced Vibration of Cylinder Arrays - Vol. 2, Paidoussis, M.P.; Chen, S.S.; Steininger, D.A.
ASME NCA - Vol. 13	Flow-Structure and Flow-Sound Interactions - Vol. 3, Farabee, T.M.; Paidoussis, M.P.
ASME PVP - Vol. 243	Acoustical Effects in FSI - Vol. 4, Paidoussis, M.P.; Sandifier, J.B.
ASME PVP - Vol. 244	Axial and Annular Flow-Induced Vibration and Instabilities - Vol. 5, Paidoussis, M.P.; Au-Yang, M.K.
ASME AMD - Vol. 151 (PVP - Vol. 244)	Fundamental Aspects of Fluid-Structure Interactions - Vol. 7, Paidoussis, M.P.; Akylas, T.; Abraham, P.B.

ASME - Performance Test Codes (PTC)

ASME PTC 4.1 - 1974	Steam generating units (R 1991)
ASME PTC 4.3 - 1988	Air heaters (R 1991)
ASME PTC 4.4 - 1981	Gas turbine heat recovery steam generators
ASME PTC 12.1 - 1978	Closed feedwater heaters (R 1987)
ASME PTC 12.1 - 1978	Closed feedwater heaters (R 1987)
ASME PTC 12.2 - 1983	Code on steam condensing apparatus (R 1988)
ASME PTC 12.3 - 1977	Deaerators (R 1990)
ASME PTC 12.4 - 1992	Moisture separator reheaters
ASME PTC 23 - 1986	Atmospheric water cooling equipment
ASME PTC 23.1 - 1983	Spray cooling systems
ASME PTC 30 - 1991	Air-cooled heat exchangers

BSI - British Standards Institute

BS 1113: 1992   Design and manufacture of water-tube steam generating plant (including superheaters, reheaters and steel tube economizers)
BS 2790: 1992   Design and manufacture of shell boilers of welded construction
BS 3274: 1960   Tubular Heat Exchangers for General Purposes
BS 3606: 1992   Specification for Steel Tubes for Heat Exchangers
BS 5500: 1997   Unfired Fusion Welded Pressure Vessels

BS 4485 - Cooling Towers
BS 4485: Part 1 - 1969   Glossary of Terms
BS 4485: Part 2 - 1988   Methods of Testing and Acceptance Testing
BS 4485: Part 3 - 1988   Thermal and Functional Design of Cooling Towers
BS 4485: Part 4 - 1996   Structural Design of Cooling Towers

CTI - Cooling Tower Institute

ATC-105: 1982   Acceptance Tests on Water Cooling Towers

HEDH - Heat Exchanger Design Handbook - 1983 (Rev. 1/85)

Vol. 1   Heat Exchanger Theory
Vol. 2   Fluid Mechanics and Heat Transfer
Vol. 3   Thermal and Hydraulic Design of Heat Exchangers
Vol. 4   Mechanical Design of Heat Exchangers
Vol. 5   Physical Properties

Hemisphere Publishing Corporation, Washington - New York - London, VDI-Verlag, Düsseldorf.

Heat Exchange Institute

Standards for Closed Feedwater Heaters, 5th Edition 1992
Standards for Steam Surface Condensers, 8th Edition 1984
Standards for Power Plant Heat Exchangers, 2nd Edition 1990
Standards for Direct Contact Barometric and Low-Level Condensers, 5th Edition 1970
Standards for Typical Specifications for Deaerators, 1st Edition, 1992

ISO - International Standardization Organization

ISO 5730: 1992   Stationary shell boilers of welded construction (other than water-tube boilers)

TEMA - Tubular Exchangers Manufacturers Association

TEMA Standards - Edition 1988

WRC - Welding Research Council

WRC 107   Local stresses in cylindrical shells due to external loadings, Edition 1979
WRC 372   Guidelines for flow-induced vibration in heat exchangers, J.B. Sandifier, 1992

Miscellaneous

HTH - 586   Heat Transfer Handbook, Design and Application of Paraflow Plate Heat Exchangers, 4th Edition, APV Crepaco Inc.

AD-Merkblätter - Gesamtausgabe 1996

DIN - Deutsches Institut für Normung

DIN 28 182 - Mai 1987	Rohrbündel-Wärmeaustauscher, Rohrteilungen, Durchmesser der Bohrungen in Rohrboden, Umlenksegmenten und Stutzplatten
DIN 28 183 - Mai 1988	Rohrbündel-Wärmeaustauscher, Benennungen
DIN 28 184, Teil 1 - Mai 1988	Rohrbündel-Wärmeaustauscher mit zwei festen Böden
DIN 28 184, Teil 2 - Mai 1988	Rohrbündel-Wärmeaustauscher mit zwei festen Böden
DIN 28 184, Teil 4 - Mai 1988	Rohrbündel-Wärmeaustauscher mit zwei festen Böden
DIN 28 185 - Mai 1988	Rohrbündel-Wärmeaustauscher, Rohrbündel-Einbauten
DIN 28 190 - April 1981	Rohrbündel-Wärmeaustauscher mit geschweißtem Schwimmkopf
DIN 28 191 - April 1981	Rohrbündel-Wärmeaustauscher mit geflanschtem Schwimmkopf

DECHEMA - Deutsche Gesellschaft für chem. Apparatewesen e.V.

DECHEMA - Monographie Band 87   Wärmeaustauscher: Konstruktion, Berechnung, Werkstoffe, Ausgabe 1980

FDBR - Fachverband Dampfkessel, Behalter- und Rohrleitungsbau e.V.

FDBR - Handbuch	Wärme- und Strömungstechnik
FDBR - Handbuch	Methoden der Festigkeitsberechnung auf Grundlage des RKF, Band 1 - 4, 1995
FDBR-Fachbuchreihe Band 2	Wärme- und Stoffübertragung in Dampferzeugern und Wärmeaustauschern, Brandt, F., Vulkan-Verlag - 2. Auflage 1995
FDBR-Fachbuchreihe Band 3	Dampferzeuger. Kesselsysteme, Energiebilanz, Strömungstechnik, Vulkan-Verlag - 1992
FDBR-Fachbuchreihe Band 5	Wärmetauscher. Aktuelle Probleme der Konstroktion und Berechnung, Podhorsky, M.; Krips, H., Vulkan-Verlag 1990

Jahrbuch der Dampferzeugungstechnik

4. Ausgabe 1980/1981
5. Ausgabe 1985/1986
6. Ausgabe 1989
7. Ausgabe 1992

Herausgegeben unter Mitwirkung der
VGB   Technische Vereinigung der Großkraftwerksbetreiber e.V., Essen und des
FDBR   Fachverband Dampfkessel, Behälter- und Rohrleitungsbau e.V.

Vulkan-Verlag, Essen.

KTA - Kerntechnischer Ausschuß

KTA 3201.2   Komponenten des Primärkreises von Leichtwasserreaktoren, Teil 2: Auslegung, Konstroktion und Berechnung, Fassung 6/96

KTA 3211.2   Druck- und aktivitätsführende Komponenten außerhalb des Primärkreises, Teil 2: Auslegung, Konstruktion und Berechnung, Fassung 6/92

TRD - Technische Regeln für Dampfkessel - Gesamtausgabe 1996

VDI - Verein Deutscher Ingenieure

VDI-Wärmeatlas   Berechnungsblätter für den Wärmeübergang 7. Auflage 1994
VDI 2047   Kühltürme - Begriffe und Definitionen - 1992

Abdazic, E. and Süßmann, W., Speisewasservorwärmer in gewickelter Ausführung - Gegenüberstellung mit anderen Bauarten, Auslegung, Betriebserfahrungen, VGB-Kraftwerkstechnik 58 (1978), Nr. 8, S. 570–574.

Alt, W., Auslegung und Betrieb von Hybridkühltürmen, VGB Kraftwerkstechnik 67 (1987), Nr. 1, S. 37–43.

Alt, M. and Alt, W., Aufbau, Wirkungsweise und Betriebserfahrungen des Naß-/Trockenkühlturms, VGB Kraftwerkstechnik 67 (1987), Nr. 4, S. 388–392.

Alt, W. and Mäule, R., Hybridkühlturm im wirtschaftlichen Vergleich zu Naß - und Trockenkuhltürmen, VGB Kraftwerkstechnik 67 (1987), Nr. 8, S. 763–768.

Andjelić, M., Stabilitätsverhalten querangeströmter Rohrbündel mit versetzter Dreiecksteilung, Dissertation Universität Hannover 1988.

Augustin, W. and Bohnet, M., Modellierung des Fouling-Verhaltens von rauhen Wärmetauscherrohren, Chemie-Ingenieur-Technik 66 (1994), Nr. 10, S. 1396–1399.

Avallone, E. A. and Baumeister, T., II, Mark's Standard Handbook for Mechanical Engineers, 10. Ausgabe 1996, McGraw-Hill.

Baehr, H. D. and Stephan, K., Wärme- und Stoffübertragung, Springer-Verlag, 1994.

Bakay, A., Der Kühlturm ohne Dampfschwaden, Brennstoff-Wärme-Kraft (BWK), 25 (1973), S. 52–54.

Bassiouny, M. K., Experimentelle und theoretische Untersuchungen über Mengenstromverteilung, Druckverlust und Wärmeübertragung in Plattenwärmeaustauschern, Fortschritt-Berichte der VDI- Zeitschriften, Reihe 6, Nr. 181, VDI-Verlag, Düsseldorf, 1985.

Bansal, B. and Müller-Steinhagen, H., Crystallization fouling in plate heat exchangers, Transactions of the ASME, J. of Heat Transfer, Vol. 115 (1993), Nr. 3, S. 584–591.

Bartz, J. A., Dry cooling of power plants — a mature technology? Power Engineering, Vol. 92 (1988), S. 25–27.

Bartz, J. A., New developments in cooling systems, Power Engineering, Vol. 95 (1991), Nr. 2, S. 29–31.

Bechmann, F., Das instationäre und stationäre Verhalten von Platten-Wärmeübertragern im industriellen Maßstab, VDI Fortschritt-Berichte, Reihe 3: Verfahrenstechnik, Nr. 437, 1996.

Becker, B. R., Effect of drift eliminator design on cooling tower performance, Transactions of the ASME, J. Eng. Gas Turbines and Power, Vol. 114 (1992), Nr. 4, S. 632–642.

Becker, N. and Renz, U., Der Einfluß der Rippenrohranordnungen auf die Abmessungen von Naturzug-Trockenkühltürmen, VGB Kraftwerkstechnik 62 (1982), Nr. 2, S. 113–119.

Becker-Balfanz, C. D., Hopp, W. -W., Königsdorf, W., Maier, K. H., and Pletka, H. D., Erfahrungen mit Platten- und Spiralwärmeübertragern, GASWÄRME International, 45 (1996), Nr. 6, S. 276–284.

Bell, R. J. and Strauss, S. D., Advancing heat-exchanger reliability, Special Report, Power 135 (1991), Nr. 7, S. 13–22.

Bischoff, M., Simulation von Gasströmungen mit Temperatur- und Geschwin-digkeitsschieflagen durch Rohrbündel, Fortschritt-Berichte der VDI, Reihe 19: Wärmetechnik/Kältetechnik, Nr. 64, VDI- Verlag 1994.

Bohnet, M., Fouling von Wärmeübertragungsflächen, Chemie-Ingenieur-Technik 57, (1985), Nr. 1, S. 24–36.

Bohnet, M., Bott, T. R., Karabelas, H. J., Pilavachi, P. A., Séméria, R., and Vidil, R., Fouling mechanisms. Theoretical and practical aspects, EUROTHERM Seminar 23, Proceedings.

Borenstein, S. W., Microbially influenced corrosion handbook, Industrial Press Inc., 1994.

Boyer, J. and Trumpfheller, G., Specification tips to maximize heat transfer, Chemical Engineering, Vol. 100 (1993), H. 5, S. 90–97.

Brab, H., Entwicklung und Anwendung eines zweidimensionalen Rechenmodells zur Untersuchung des transienten Verhaltens eines Röhrenspaltofens mit Segment-Umlenkblechen, Dissertation Technische Hochschule Aachen, 1987.

Brauer, H., Strömungswiderstand und Wärmeübergang bei quer angeströmten Wärmeaustauschern mit kreuzgitterförmig angeordneten glatten und berippten Rohren, Chemie-Ingenieur-Technik 36, (1964), Nr. 3, S. 247–260.

Brauer, H. and Kim, T.-H., Strömung um Rohrreihen und durch Rohrbündel sowie Wärmeübergang im Bereich niedriger Werte der Reynoldszahl, Teil 1: Eine Reihe mit fluchtender Anordnung der Rohre, Forschung im Ingenieurwesen 59 (1993), Nr. 7/8, S. 129–153.

Briem, K. and Naujoks, J. H. : Wärmerückgewinnung in Hoch- und Niedertemperaturprozessen, Gas-Wärme-International 34, (1985), Nr. 9, S. 362–370.

Buehler, J. W., Sikes, R. K., Kuritsky, J. N. et al., Prevalence of antibodies to legionella pneumophila among workers exposed to contaminated cooling tower, Archives of Environmental Health 40 (1985), S. 207–210.

Bullerschen, K. G. and Wilhelmi, H., Kühlung von Lichtbogenofenelektroden durch Wärmerohre, Stahl und Eisen 110 (1990), Nr. 8, S. 91–98.

Burger, R., Cooling towers. The often over-looked profit center, Chemical Engineering, Vol. 100 (1993), H. 5, S. 100–104.

Burger, R., Cooling Tower Technology: Maintenance, Upgrading and Rebuilding, Burger & Associates Inc., 3. Ausgabe 1996.

Burmester, H. and Zwahr, H., Rauchgaswiederaufheizung für SCR-Anlage nach REA, Brennstoff-Wärme-Kraft (BWK), 38 (1986), Nr. 5, S. 206–210.

Buuck, M., Wärmetransportleistung rotierender Wärmerohre in Luftkanälen, Fortschritt-Berichte VDI, Reihe 19: Wärmetechnik/Kältetechnik, Nr. 47, VDI-Verlag 1991.

Buxmann, J. and Johannsen, T., Beitrag zur Auslegung von titanberohrten Kondensatoren unter Berücksichtigung schwingungserregender Mechanismen, Fortschritt-Berichte VDI, Reihe 6: Energieerzeugung, Nr. 215, VDI-Verlag, Düsseldorf, 1988.

Capitaine, D., Jentsch, W., and Stoffels, P. H., Der Einsatz von Abhitzekesseln und einige Konstruktionsmerkmale, VGB-Mitteilungen 49, (1969), Nr. 3, S. 165–173.

Chen, S. S., Flow-Induced Vibration of Circular Cylindrical Structures, Hemisphere Publishing Corporation, Washington - New York - London, 1987.

Chenoweth, J. M., Kaellis, J., Michel, J. W., and Shenkman, S., Advances in Enhanced Heat Transfer, 1979, ASME, New York.

Cheremisinoff, N. P. and Cheremisinoff, P. N., Heat Transfer Equipment, Prentice Hall, Inc., 1993.

Chisholm, D., Developments in Heat Exchanger Technology, Applied Science Publishers Ltd., London, 1980.

Chuse, R. and Carson, Br. E., Sr., The ASME Code Simplified - Pressure Vessels, 7. Ausgabe 1993, Donelley & Sons Company.

Danis, J. I., Material Selection, Fabrication and Inspection of Refinery Heat Exchangers, Welding Journal, Vol. 65, (1986), Nr. 6, S. 25–30.

Dowling, N. J., Mittelmann, W., and Danko, J. C., Microbially influenced corrosion and biodeterioration, NACE - National Association for Corrosion Engineers, 1991.

Eck, B., Technische Strömungslehre, 5. Auflage 1957, Springer-Verlag.

Eckels, P. W. and Rabas, T. J., Heat Transfer and Pressure Drop Performance of Finned Tube Bundles, Transactions of the ASME, Journal of Heat Transfer, Vol. 107, (1985), Nr. 1, S. 205–213.

Eckert, E. R. G., Einführung in den Wärme- und Stoffaustausch - 3. Auflage 1966, Springer-Verlag.

Effertz, P. H., Forchhammer, P., and Heinz, A., Korrosion und Erosion in Speisewasservorwärmern, Der Maschinenschaden 51, (1978), Nr. 4, S. 154-161.

Effertz, P. -H., Spröde Rohrreißer in den Verdampferrohren von Naturumlaufkesseln nach ungewöhnlich heißer Heißwasseroxidation, VGB Kraftwerkstechnik (70) 1990, Nr. 1, S. 53–59.

Eggert, H., Der vollverschweißte Plattenwärmetauscher als Heizungs- und Niederdruckvorwärmer im Kraftwerk, VGB Kraftwerkstechnik (75) 1995, Nr. 5, S. 433–435.

Eimer, K. and Besold, D., Leistungsverluste von Kraftwerken durch "biologische Verschmutzung" der Kühlwassersysteme und umweltfreundliche Gegenmaßnahmen, VGB Kraftwerkstechnik 68 (1988), Heft 6, S. 610–616.

Eisenbeis, H., Oleas, J., Hosp, A., and Horlacher, H., Einsatz von Hochgeschwindigkeitsabscheidern (MOPS/SRUPS) im Kernkraftwerk Beznau I, VGB-Kraftwerkstechnik 74 (1994), Nr. 12, S. 1055–1060.

Eisinger, F. L., Unusual acoustic vibration of a shell and tube process heat exchanger, Transactions of the ASME, Journal of Pressure Vessel Technology, Vol. 116 (1994), Nr. 2, S. 141–149.

Elliot, T. C., Cooling towers—Special Report, Power, Vol. 129, (1985), Nr. 12, S. 1-16.

Elliot, T. C. and Kals, W., Air-cooled condensers—Special Report, Power, Vol. 134 (1990), S. 13–21.

Elsner, N. and Dittmann, A., Grundlagen der technischen Thermodynamik, Band 1: Energielehre und Stoffverhalten, 8. Auflage 1992, Akademie Verlag.

Elsner, N., Fischer, S., and Huhn, J., Grundlagen der technischen Thermodynamik, Band 2: Wärmeübertragung, 8. Auflage 1993, Akademie Verlag.

Epstein, N., Thinking about Heat Transfer Fouling: A 5×5 Matrix, Heat Transfer Engineering, Vol. 4 (1983), Nr. 1, S. 43–55.

Erdmann, C., Wärmeaustauscher mit zirkulierender Wirbelschicht zur Verhinderung von Belagbildung, Fortschritt-Berichte VDI, Reihe 3, Nr. 346, VDI-Verlag 1993.

Ernst, G. et al., Hybridkühlturm. Betriebsverhalten und Schwadenausbreitung, VGB-Kraftwerkstechnik 64 (1984), Nr. 10, S. 918–923.

Fago, B., Wirbelresonanzanregung von Kreis- und Quadratzylindern, Fortschritt-Berichte VDI, Reihe 7: Strömungstechnik, Nr. 263.

Faghri, A., Heat Pipe Science and Technology, Taylor & Francis, 1995.

Filleböck, A., Tschaffon, H., Gerhards, U., and Akermann, P., Erfahrungen aus einer Versuchsanlage zur Wiederaufheizung von Rauchgasen (STAGAVO) in einer Braunkohle-REA, VGB-Kraftwerkstechnik 75 (1995), Nr. 10, S. 885–888.

Flemming, H.-C., Biofilme und Wassertechnologie, Teil I: Entstehung, Autbau, Zusammensetzung und Eigenschaften von Biofilmen, GWF-Wasser-Abwasser 132. Jahrgang (1991), Nr. 4, S. 197–207.

Flemming, H.-C., dito; Teil II: Unerwünschte Biofilme-Phänomene und Mechanismen, GWF-Wasser-Abwasser, 133. Jahrgang (1992), Nr. 3, S. 113–127.

Flemming, H.-C., dito; Teil III: Bekämpfung unerwünschter Biofilme, GWF-Wasser-Abwasser 133. Jahrgang (1992), Nr. 6, S. 298–310.

Flemming, H.-C., Mikrobielle Werkstoffzerstörung - Grundlagen: Ökonomischtechnischer Überblick, Werkstoffe und Korrosion 45 (1994), Heft 1, S. 5–9.

Flemming, H.-C. and Schaule, G., Mikrobielle Werkstoffzerstörung—Biofilm und Biofouling: Biofouling, Werkstoffe und Korrosion 45 (1994), S. 29–39.

Flemming, H.-C., Biofouling und Biokorrosion—die Folgen unerwünschter Biofilme, Chemie Ingenieur Technik 67 (1995), Heft 11, S. 1425–1430.

Focke, W. F., Selecting optimum plate heat exchanger surface patterns, Transactions of the ASME, Journal of Heat Transfer, Vol. 108 (1986), Nr. 1, S. 153–166.

Franklin, H. N., Roper, D. R., and Thomas, D. R., Internal Bore Welding Method Repairs Condenser Leaks, Welding Journal, Vol. 65 (1986), Nr. 12, S. 49–52.

Franklin, H. N., Building a better heat pipe, Mechanical Engineering, Vol. 112, 1990, Nr. 8, S. 52–54.

Fraser, D. W., Deubner, D. C., and Gilliam, D. K., Nonpneumonic, short-incubation period legionellosis (Pontiac fever) in men who cleaned a steam turbine condenser, Science, Vol. 205, 17. August 1979, S. 690–691.

Frauenfeld, M., Ljungström-Gasvorwärmer zur Wiederaufheizang naßentschwefelter Reingase, VGB Kraftwerkstechnik 63 (1983), Nr. 3, S. 229–232.

Gasteiger, G., Ein Beitrag zur Ermittlung fluidelastischer Koppelschwingungen in Rohrbündelwärmeaustauschern, Fortschritt-Berichte der VDI-Zeitschriften, Reihe 6, Nr. 124, VDI-Verlag, Düsseldorf, 1983.

Granser, D., Hosenfeld, H.-G., and Schwerdtner, O.-A., Strömungsuntersuchungen zur Entwicklung großer Dampfturbinenkondensatoren, Brennstoff-Wärme-Kraft (BWK) 37 (1985), Nr. 10, S. 397–406.

Gregorig, R., Wärmeaustauscher - Ausgabe 1959, Verlag H.R. Sauerländer.

Gregorig, R., Einige Sonderprobleme beim Entwurf der Wärmeaustauscher mit Phasenänderung, Chemie-Ingenieur-Technik 36 (1964), Nr. 3, S. 261–268.

Groß, H.-G., Untersuchung aeroelastischer Schwingungsmechanismen und deren Berücksichtigung bei der Auslegung von Rohrbündelwärmeaustauschern, Dissertation Universität Hannover, 1975.

Guyer, E. C. and Bartz, J. A., Dry cooling moves into the main stream, Power Engineering, Vol. 95 (1991), S. 29–31.

Gvodenac, D. D., Experimental prediction of heat transfer coefficients by use of a double-blow method, Wärme- und Stoffübertragung 29 (1994), Nr. 6, S. 361–365.

Halle, H., Chenoweth, J. M., and Wambspanss, M. W., Shellside Waterflow-Induced Vibration in Heat Exchanger Configurations with Tube Pitch-to-Diameter Ratio of 1.42, Transactions of the ASME, J. of Pressure Vessel Technology, Vol. 111, November 1989 (4), S. 441–449.

Heider, W., Entwicklung, Konstruktion und Test eines SiSiC-Rohrbündelwärmeaustauschers, S. 52–59, Handbuch Technische Keramik, 2. Ausgabe 1990, Vulkan-Verlag Essen.

Heinrich, J., Huber, J., Schalter, H., Ganz, R., and Heinz, O., Energieeinsparung mit Hilfe keramischer Wärmeaustauscher, GASWÄRME International 40 (1991), Nr. 5, S. 199–206.

Heitz, E., Mikrobielle Werkstoffzerstörung—Grundlagen: Grundvorgänge der Korrosion, Werkstoffe und Korrosion 45 (1994), S. 17–20.

Held, H.-D. and Bohnsack, G., Kühlwasser—Verfahrenstechnische und chemische Methoden der Kühlwasserbehandlung in Industrie und Kraftwerken - Süßwasser - Meerwasser - Brackwasser, 3. Auflage 1984, Vulkan-Verlag.

Henning, H., Stand und Entwicklung im Kühlturmbau, Technische Mitteilungen 78 (1985), Nr. 10, S. 511–524.

Hess, F. and Thier, B., Apparate—Bau, Technik, Anwendung, Handbuch 1. Ausgabe 1990, Vulkan-Verlag.

Hewitt, G. F., Shires, G. L., and Bott, T. R., Process Heat Transfer, CRC Press, Inc., 1994.

Hinchley, P., Prozeßgaskühler—Bauarten und Schadensfälle, Chemie-Ingenieur-Technik, 49 (1977), Nr. 7, S. 553–557.

Hirschfelder, G., Der Trockenkühlturm des 300-MW-THTR-Kernkraftwerks Schmehausen-Uentrop, VGB Kraftwerkstechnik 53 (1973), Nr. 7, S 463–471.

Hobson, E., Lindahl, P., and Massey, T., Leistungssteigerung mit Kühlturmeinbauten aus NPF (National Power Fill), VGB Kraftwerkstechnik 75 (1995), Nr. 9, S. 829–832.

Honekamp, H. and Katzmann, A., Zur neueren Entwicklung von Kühltürmen, Technische Mitteilungen 74 (1981), Nr. 7, S. 391–394.

Honekamp, H. and Katzmann, A., Kühlturmeinbauten aus Keramik, VGB-Kraftwerkstechnik 66 (1986), Nr. 5, S. 130–132.

Huttner, F. and Winkler, R., Belagbildung auf Wärmeubertragungsflächen—eine kritische Literaturauswertung, Energietechnik 36 (1986), Nr. 4, S. 147–151.

Issler, L., Ruoß, H., and Häfele, P., Festigkeitslehre—Grundlagen, Springer-Verlag, 1995.

Joo, Y. and Dhir, V. K., On the mechanism of fluidelastic instability of a tube placed in an array subjected to two-phase crossflow, Transactions of the ASME, J. of Fluids Engineering 117 (1995), Nr. 4, S. 706–712.

Junker, A., Cooling Towers—Kühltürme, TexTerm, VCH Verlagsgesellschaft 1991.

Juran, H. and Kiene, K., Regenerativer Wärmeaustausch bei tiefen und hohen Temperaturen, Brennstoff-Wärme-Kraft (BWK) 42 (1990), Nr. 5, S. 261–264.

Juran, H. and Plocki, O., Das neue Kühlturmprinzip Matrix-Multiflow, Brennstoff-Warme-Kraft (BWK), 47 (1995), Nr. 11, S. 480–484.

Juran, H., Wärmeverschiebungs- und Entsorgungssysteme. Teil 1: Übersicht und Anwendung, 3R International, 27 (1988), Nr. 2, S. 110–116.

Juran, H., Wärmeverschiebungs- und Entsorgungssysteme. Teil 2: Anwendungstechnische Dynamisierung durch Umweltgesetzgebung, 3R International, 28 (1989), Nr. 7, S. 436–441.

Kanthak, R., Hochdruckvorwärmer (HDV) moderner Konstruktion für konventionelle Kraftwerke, Mitteilungen aus dem Kraftwerksanlagenbau, 27 (1987), Nr. 2, S. 12–15.

Kapsa, M., Experimentelle und theoretische Untersuchung zur Beeinflussung von ungleichförmigen Temperatur- und Geschindigkeitsprofilen durch Rohrbündel, Fortschritt-Berichte VDI, Reihe 6: Energierzeugung, Nr. 298, VDI-Verlag.

Kellenbeck, J., Wärmeübergang bei der Kondensation von strömenden Dämpfen reiner Stoffe und binärer Gemische, Fortschritt-Berichte VDI- Reihe 3, Nr. 365, VDI-Verlag 1994.

Kelp, F., Welch, R., and Charalambus, B., Entwicklung des Duplex-Geradrohrvorwärmers für große Dampfturbosätze, VGB-Kraftwerkstechnik 56 (1976), Nr. 8, S. 489–496.

Kim, N.-H. and Webb, R. L., Analytic Prediction of the Friction and Heat Transfer for Turbulent Flow in Axial Internal Fin Tubes, Transactions of the ASME, J. of Heat Transfer, Vol. 115 (1993), Nr. 3, S. 553–559.

Kim, W.-K., Wärmeübergang und Druckverlust in längsdurchströmten Rohrbündel-wärmeübertragern, Fortschrin-Berichte VDI, Reihe 19, Nr. 80, VDI-Verlag 1994.

King, R., Flow-Induced Vibrations, Proceedings of the 1st International Conference, Bowness-on-Windermere, England, 12–14 Mai 1987, BHRA - The Fluid Engineering Centre, UK, Springer-Verlag 1987.

Klaren, D. G., Non-fouling Wirbelbett-Wärmetauscher für Schmierölanlagen, Erdöl und Kohle 42 (1989), Nr. 11, S. 417–419.

Kotter, M., Lintz, H. -G., and Turek, T., Katalytische Stickoxidreduktion in einem rotierenden Wärmeübertrager, Chemie-Ingenieur-Technik, 64 (1992), Nr. 2, S. 446–448.

Krause, S., Neue Untersuchungen zum Fouling von Wärmeübertragungsflächen durch Sedimentbildung und Kristallisation, Vortrag auf dem Jahrestreffen der Verfahrens-Ingenieure, 25–27 Sept. 1985, Hamburg, Chemie-Ingenieur-Technik, MS 1447/86.

Krause, S., Fouling an Wärmeübertragungsflächen durch Kristallisation und Sedimentbildung, VDI- Forschungsheft 637, VDI-Verlag, Düsseldorf, 1986.

Kritzler, G. and Kraft, E., Entwicklung im Bau von Luftvorwärmern, Technische Mitteilungen, 65 (1972), Nr. 2, S. 76–82.

Kritzler, G. and Kraft, E., Beitrag zum Entwicklungsstand großer Regenerativ-Lufivorwärmer mit feststehender Heizfläche (Bauart Rothemühle), VGB-Kraftwerkstechnik 65 (1985), Nr. 7, S. 677–683.

Kröger, D. G. : Thermische Strömung in trockengekühlten Systemen für Kraftwerke, VGB Kraftwerkstechnik 71 (1991), Nr. 11, S. 1013–1016.

Kukral, R., Modelle zur Beschreibung der Zustandsänderungen in Rohrbündel-Wärmeübertragern bei zeitlich veränderlichen Betriebsbedingungen, Fortschritt-Berichte VDI, Reihe 19, Nr. 76, VDI-Verlag 1994.

Kummel, J., Abhitze- und Sonderkessel in der chemischen und petrochemischen Industrie, Chemie-Ingenieur-Technik 49 (1977), Nr. 6, S. 475–479.

Lange, A., Kosteneinsparungen durch verbesserten Betrieb der Kühlrohr-Reinigungsanlage, Einsatz von Kühlwasserfiltern und einer neuartigen Kondensatorüberwachung, VGB Kraftwerkstechnik 79 (1990), Nr. 8, S. 681–688.

Lavis, G., Evaporators—How to make the right choice, Chemical Engineering, Vol. 101 (1994), Nr. 4, S. 92–102.

Lee, D.-Y. W., Thermisches Verhalten von Rohrbündelwärmeübertragern, VDI, Fortschritt-Reihe 19, Nr. 18, VDI-Verlag 1994.

Lewin, G., Lässig, G., and Woywode, N., Apparate und Behälter—Grundlagen Festigkeitsberechnung, VEB Verlag Technik 1990.

Leyh, T., Jahr, M., and Gelke, H., Strömungsinduzierte Rohrbündelschwingungen in Wärmeübertragern, VGB-TB 231, Forschung in der Kraftwerkstechnik, 1993.

Marner, W. J., Gas-side fouling, Mechanical Engineering, Vol. 108 (1986), Nr. 3, S. 70–77.

Marto, P. J. and Kroeger, P. G., Condensation Heat Transfer, 1979, ASME, New York.

Mayinger, F., Sieden—Stabilisator und Störfaktor sicheren Betriebs, Chemie-Ingenieur-Technik 56 (1984), Nr. 3, S. 169–179.

McAdams, W. H., Heat Transmission, 3rd Edition, 1954, McGraw-Hill Book Company Inc., New York.

McDade, J. E., Shephard, C. C., Fraser, D. W. et al., Legionnaires disease, The New England Journal of Medicine, Vol. 297, Nr. 22, 1977, S. 1197–1203.

Mehra, D. K., Shell-and-tube heat exchangers, Chemical Engineering, 25. Juli 1983, S. 47–56.

Merker, G. P. and Hanke, H., Druckverlust und Stoffübergang in querangeströmten kompakten Ovalrohrbündeln, Chemie-Ingenieur-Technik, MS 1309/1985.

Meier, F., Struberg, K., Tegetoff, H., Trobitz, M., and Thoma, K., Sanierung von Wärmetauschern—Teilrohraustausch und Sleeven, Atomwirtschaft-Atomtechnik (ATW), 42 (1997), S. 30–33.

Meyer, K., Probleme der Entwicklung großer Oberflächenvorwärmer, Energietechnik 40 (1990), Nr. 8, S. 289–292.

Miller, C. W., Heutiger Stand der Auslegung und Fertigung von Vorwärmern, VGB-Kraftwerkstechnik 64 (1984), Nr. 11, S. 982–989.

Minton, P. E., Handbook of Evaporation Technology, 1986, Noyes Publications, Park Ridge, New Jersey, USA.

Mitterecker, E. and Kallenberg, H., Speisewasservorwärmer großer Dampfkraftwerke, Brennstoff-Wärme-Kraft (BWK) 37 (1985), Nr. 10, S. 388–396.

Moretti, P. M., The Paradox of Flow-Induced Vibrations, Mechanical Engineering, Vol. 105 (1986), Nr. 12, S. 56–61.

Müller-Steinhagen, H. and Reif, F., Thermische und hydrodynamische Einflüsse auf die Ablagerung suspendierter Partikeln an beheizten Flächen, VDI- Fortschritt-Berichte, Reihe 19: Wärmetechnik/Kältetechnik, Nr. 40.

Nasser, O., Hybrid-Wärmeübertrager. Euroheat & Power—Fernwärme International, 25 (1996), Nr. 10, S. 560–563.

Pahl, H. and Muschelknautz, E., Statische Mischer und ihre Anwendung, Chemie-Ingenieur-Technik 52 (1980), S. 285–291.

Paikert, P., Wärmerohre in der industriellen Praxis CChemie-Ingenieur-Technik 62 (1990), Nr. 4, S. 278–286.

Paikert, P., Wärmeruckgewinnung mit KV-Systemen und mit Wärmerohren, Technische Mitteilungen 83 (1990), Heft 2, August, S. 101–115.

Pietsch, M., Spielmann, M., and Werner, H. P., Vorkommen von Legionellen in Wasserkreisläufen von Rückkühlanlagen, Hygiene und Medizin 13 (1988), S. 229–232.

Poredooš, A. and Gašperšič, B., Temperaturverteilung und Wirkungsgrad einer trockenen quadratischen Rippe, Forschung im Ingenieurwesen 59 (1993), Nr. 6, S. 105–109.

Puckorius, P. R., Cooling-water treatment, Special Report, Power, Vol. 139 (1995), Nr. 5, S. 17–28.

Rathje, U. J. and Pflaumbaum, H.-J., Die Generation 2000 luftgekühlter Abdampfkondensatoren, VGB-Kraftwerkstechnik 76 (1996), Nr. 1, S. 31–36.

Reid, D. D. and Taborek, J., Selection criteria for plain and segmented finned tubes for heat recovery systems, Transactions of the ASME, Journal of Engineering for Gas Turbines and Power, Vol. 116 (1994), Nr. 2, S. 407–410.

Remberg, H.-W. and Fehndrich, B., Wirkungsgradverbesserung bei Naturzug-Naßkühltürmen durch Austausch von Asbestzement-Kühleinbauten gegen Kunststoff-Kühleinbauten, VGB-TB 313 Kraftwerk und Umwelt 1993, S. 112–117.

Remberg, H.-W., Rauchgasableitung über Naturzugkühltürme—Umrüstungsmaßnahmen und Betriebserfahrungen, Energietechnik, 42. Jg., Nr. 5, 1992, S. 177–181.

Rennhack, R. and Numrich, R., Die Auslegung von Kühlerkondensatoren zur partiellen Kondensation von Dämpfen aus strömenden Gas/Dampf-Gemischen, Chemie-Ingenieur-Technik 57 (1985), Nr. 4, S. 278–289.

Rinenhouse, R. C., Industry weapons grow in biofouling battle, Power Engineering, Vol. 95 (1991), Nr. 10, S. 17–23.

Roth, J. E. and Roetzel, W., Beulrohre mit verbessertem Wärmeübergang zum Einsatz in Wärmeaustauschern, Vortrag auf dem Jahrestreffen der Verfahrens-Ingenieure in Hamburg, 25–27 September 1985, Chemie-Ingenieur-Technik MS 1449/86.

Samdami, G., Fouhy, K., and Moore, S., Heat Exchange: The Next Wave, Chemical Engineering 100 (1993), Nr. 6, S. 30–35.

Sand, W., Mikrobielle Werkstoffzerstörung—Grundlagen: Mikrobielle Schädigungsmechanismen. Werkstoffe und Korrosion 45 (1994), Heft 1, S. 10–16.

Sarma, P. K. and Saibabu, J., Evaporation of laminar, falling liquid film on a horizontal cylinder, Wärme- und Stoffübertragung 27 (1992), Nr. 6, S. 347–355.

Sattler, K., Thermische Trennverfahren—Grundlagen, Auslegung, Apparate, 2. Auflage 1995, VCH Verlagspesellschaft GmbH.

Schack, A., Der industrielle Wärmeübergang, 5. Auflage 1957, Verlag Stahleisen.

Schmitz, H.-P., Dictionary of Pressure Vessel and Piping Technology, FDBR-Fachwörterbuch Band 1 und 2, Vulkan-Verlag 1991.

Schneider, S., Lokaler Wärmeübergang im Außenraum von längsangeströmten Rohrbündelapparaten, Chemische Technik 45 (1993), Nr. 5, S. 392–400.

Schneider, J. and Wimmler, B., Effizienzsteigerung fossil befeuerter Kraftwerke durch Verbesserung des "Kalten Endes"—ein Beitrag zur Umweltentlastung, Energieanwendung/Energie- und Umwelttechnik 42. Jahrgang (1993), Nr. 12, S. 649–652.

Schnell, H. and Thier, B., Wärmeaustauscher—Energieeinsparung durch Optimierung von Wärmeprozessen, Vulkan-Verlag, 1. Ausgabe 1991.

Schnell, H. and Thier, B., Wärmeaustauscher—Energieeinsparung durch Optimierung von Wärmeprozessen, Vulkan-Verlag, 2. Ausgabe 1994.

Schroder, K., Große Dampfkraftwerke—Planung, Ausführung und Bau, DriRer Band, Die Kraftwerksausrüstung: hier: Heyde, H., Apparate im Wasser/Dampfkreislauf, I: Wärmetauscher, S. 276–351.

Schult, M., Untersuchungen zum dynamischen Verhalten von Doppelrohrwärmeaustauschern mit doppeltem Phasenwechsel, Dissertation Universität Hannover, 1979.

Schumacher, A. and Waldmann, H., Wärme- und Strömungstechnik im Dampferzeugerbau, MAN - Sonderdruck 1972, Vulkan-Verlag, Essen.

Schwaigerer, S., Festigkeitsberechnung im Dampfkessel, Behälter- und Rohrleitungsbau, 4. Auflage 1983, Springer-Verlag.

Schweisfurth, R., Mikrobiologie der Kühlkreisläufe, Technische Mitteilungen, 78. Jahrgang (1985), H. 9, S. 432–437.

Schwieger, R. G., Heat Exchangers, Power Special Report, June 1970.

Seetharamu, K. N. and Swaroop, S., The effect of size on the performance of a fluidized bed cooling tower, Wärme- und Stoffübertragung, 26 (1991), Nr. 1, S. 17–21.

Shah, R. K., What's New in Heat Exchanger Design, Mechanical Engineering, Vol. 105 (1984), Nr. 5, S. 50–59.

Singer, J. G., Combustion—Fossil Power Systems, A Reference Book on Fuel Burning and Steam Generation, Third Edition 1981, Combustion Engineering, Windsor.

Sipcevič, B., Druckabfall im Mantelraum von Rohrbündel-Wärmeübertragen mit Kreisscheiben und -ringen, Technische Rundschau Sulzer, Heft 1/1978, S. 28–30.

Sohngen, R., Gestaltung und Normung von Wärmeaustauschern 1981, Bayer AG, Leverkusen.

Spence, J. R. and Ryall, A., The Development and Production of High Presssure Feed Heaters for Modern Central Power Stations, Combustion, Vol. 40 (1969), Nr. 3, S. 27–39.

Spence, J. and Tooth, A. S., Pressure Vessel Design—Concepts and principles, E & FN Spon, Imprint of Chapman Hall, 1994.

Stelzer, F., Wärmeübergang und Strömung, Thiemig-Taschenbücher Bd. 18, Verlag Karl Thiemig GK, München 1971.

Stephan, K. and Grigull, U., Wärmeübergang beim Kondensieren und beim Sieden, Springer-Verlag, 1988.

Stockmeier, D., Einfluß der Viskosität auf die Schwingungen in Rohrbündelapparaten, Dissertation Universität Hannover, 1990.

Strauss, S. D., Barrier dispersion mitigates biofouling, MIC effects, Power, Vol. 136 (1992), Nr. 3, S. 74/75.

Streng, A., Kombinierte Naß-/Trockenkühltürme in Zellenbauweise, Brennstoff-Wärme-Kraft BWWK, 47 (1995), Nr. 5, S. 218–224.

Suhr, L. and Juran, H., Abgaskühlung mit Fluorkunststoffen eröffnet riesiges Wärmenutzungspotential, Energie, Jahrgang 44 (1992), Nr. 4, S. 30–35.

Taborek, J., Rose, J., and Tanasawa, I., Condensation and Condenser Design, Proceedings of the Engineering Foundation Conference on Condensation and Condenser Design, Florida, 7–12 März 1993, ASME, New York.

Tangemann, H., Fluidelastische Schwingungen von Einzelrohren in querangestromten Rohrbündeln, Dissertation Universität Hannover, 1979.

Thiel, G., Zu Problemen der thermischen Schutzschichtversprödung in beheizten Kesselrohren, Energietechnik 40 (1990), Heft 9, Sept., S. 346–349.

Tiller, A. K. and Sequeira, C. A. C., Microbial Corrosion, Proceedings of the 3rd International EFC Workshop, European Federation of Corrosion Publications, No. 15, The Institute of Materials, 1995.

Titze, H., Elemente des Apparatebaus, 1963, Springer-Verlag.

Titze, H., Wilke, H.-P., and Groß, K., Elemente des Apparatebaus, 3. Ausgabe 1992, Springer-Verlag.

Troidl, H. and Strohmeier, K., Strömungsinduzierte Schwingungen querangeströmter Rohrfelder aus der Sicht des Konstrukteurs, VGB-Kraftwerkstechnik 67 (1987), Nr. 3, S. 291–300.

Upmalis, A., Rekuperative und regenerative Lufterhitzer, Energie 24 (1974), Nr. 10, S. 249–254.

Upmalis, A., Leistungsvermögen der Fächerrippenrohre und der Rohre mit glatten runden Rippen, Wärme 85 (1979), Nr. 1, S. 9–13.

VDMA-Arbeitskreis Kühltürme, Tagungsbericht 2. Kühlturm-Tagung vom 26.02. 1992, Dresden.

VDMA-Arbeitskreis Kühltürme, Tagungsbericht 3. Kühlturm-Tagung vom 14.02. 1995, Frankfurt.

Veser, K., Regenerativ-Luftvorwärmer an ölgefeuerten Kesseln, VGB Kraftwerkstechnik 59 (1979), Nr. 1, S. 53–58.

Veser, K., Ausführungsmöglichkeiten von Regenerativ-Luftvorwärmern für die getrennte Aufheizung von Primär- und Sekundärluft, VGB Kraftwerkstechnik 61 (1981), Nr. 7, S. 579–586.

Veser, K., Regenerativ-Wärmetauscher in der Umwelttechnik, Betriebserfahrungen mit dem Gasvorwärmer an Naßentschwefelungsanlagen, mit Pilot-Anlagen denoxgerechter Luft- und Gasvorwärmer an Entstickungsanlagen und Anordnungskriterien für solche; Entwicklungsstand des DENOX-Luvo/DENO-Gavo, VGB Kraftwerkstechnik 66 (1986), Nr. 12, S. 1123–1130.

VGB-Fachtagung Kühltürme 1991, VGB Technische Vereinigung der Großkraftwerksbetreiber.

Wagner, W., Praktische Strömungstechnik, Konus-Handbuch Bd. 2, 1976, Kommissionsverlag Technischer Verlag Resch KG, Gräfelfing.

Wagner, W., Wärmeübertrager, Konstruktionsrichtlinien, Werkstoffe, Anwendungsbeispiele, 1984, Expert-Verlag, Bd. 141, Kontakt und Studium, Maschinenbau.

Wagner, W., Wärmeträgertechnik mit organischen Medien, 5. Auflage 1994, Verlag Dr. Resch.

Wagner, W., Wärmeaustauscher—Grundlagen, Aufbau und Funktion thermischer Apparate, Vogel Fachbuch, Kamprath-Reihe, 1. Auflage 1993, Vogel-Verlag.

Wagner, W., Wärmeübertragung—Grundlagen, Vogel Fachbuch, Kamprath-Reihe, 4. Auflage 1993, Vogel-Verlag.

Wang, C., Xu, D.-Q., and Li, S.-P., Investigation of heat transfer on the Dowtherm A gravity assisted heat pipe, Wärme 2- und Stoffübertragung 29 (1993), Nr. 1, S. 1–8.

Watson, B., Modular plate-type air heaters offer leak-tight service, Power, Vol. 137 (1993), Nr. 5, S. 56–57.

Webb, R. L., Principles of Enhanced Heat Transfer, John Wiley & Sons, 1994.

Webb, R. L. and Bergles, A. E., Heat Transfer Enhancement: Second Generation Technology, Mechanical Engineering, Vol. 104 (1983), Nr. 6, S. 60–67.

Weichsel, M. and Heitmann, W., Abwärmenutzung in industriellen Prozessen, Technische Mitteilungen, 70 (1977), Nr. 5, S. 259–272.

Welch, R., Anordnung der ND-Vorwärmer in Turbinenabdampfstutzen—Schaltungen, Bauarten und Montagemöglichkeiten, VGB-Kraftwerkstechnik 55 (1975), Nr. 6, S. 360–363.

Wenzel, G. and Müller-Steinhagen, H., Unterkühltes Sieden strömender Flüssigkeitsgemische, Wärme- und Stoffübertragung 26 (1991), S. 265–271.

Werner, H.-P. and Pietsch, M., Bewertung des Infektionsrisikos durch Legionellen in Kühlkreisläufen von Kraftwerken, VGB Kraftwerkstechnik 71 (1991), Nr. 8, S. 785–787.

Wersel, M. and Ridell, B., Plattenwärmeaustauscher in der Kraftwerksindustrie, Sammelband VGB-Konferenz "Kraftwerkskomponenten 1984", S. 167–173.

Widua, J., Zum Einsatz der Seeding—Technik im Horizontalrohrverdampfer bei der Eindampfung häretbildenden, wäßriger Lösungen, VDI Fortschrittberichte, Reihe 3, Nr. 410, 1995.

Wieland, G., Wasseraufbereitungsverfahren für Kühlwasserkreisläufe, Technische Mitteilungen, 81 (1988), Nr. 2, S. 91–98.

Wieser, R., Hydraulisch gekoppelte Doppelrohrluftvorwärmanlage für gasbeheizte Dampferzeuger und Industriefeuerungen, if-Industriefeuerung, Heft 51 (1991), S. 48–52.

Winkler, R. and Middelmenne, J., Epoxidharzbeschichtungen in Kondensatoren und Kühlern, Energietechnik 40 (1990), Nr. 8, S. 307–310.

Wongwises, S., Experimentelle und theoretische Untersuchungen zum Gegenstrom von Gas und Flüssigkeit in einer Rohrleitung mit Rohrbogen, Dissertation Universität Hannover, 1994.

Woodruff, E. B., Lammers, H. B., and Lammers, T. F., Steam Plant Operation, 6th edition 1992, McGraw Hill, Inc..

Yarden, A. L. and Tam, C. W., The MSR evolution, Nuclear Engineering International, Vol. 39 (1994), S. 44–45.

Zafeiriou, E. and Wurz, D., Numerische Simulation der Wärmeübertragungsvorgänge in rotierenden Regeneratoren, VGB Kraftwerkstechnik 76 (1996), Nr. 6, S. 463–470.

Zhou, X. and Bier, K., Zum Einfluß des Wandmaterials auf den Wärmeübergang beim Blasensieden in freier Konvektion, VDI Fortschritt-Berichte, Reihe 3: Verfahrenstechnik, Nr. 459, 1996.

Zimmermann, P. and Hamers, J. P., Planung und Bau eines schwadenfreien und gerauscharmen Hybridkühlturms mit einer umweltschonenden Wasserbehandlung durch

Ozon für ein GuD-Kraftwerk in den Niederlanden, VGB Kraftwerkstechnik 76 (1996), Nr. 6, S. 502–505.

Zukauskas, A., Ulinskas, R., Katinas, V., and Karni, J., Fluid dynamics and flow-induced vibrations of tube banks, Hemisphere Publishing Corporation, 1988.

TOPICAL REFERENCES

Dished Ends

1. Findlay, G. E., Moffat, D. G., and Stanley, P., Elastic Stresses in Torispherical Drumheads: Experimental Verification, *J. Strain Anal.*, 3(3) (1968), 214–225.
2. Shield, R. T. and Drucker, D. C., Design of Thin Walled Torispherical and Toriconical Pressure Vessel Heads, *J. Applied Mech.*, Trans. ASME (1961) (83), 292–297.
3. Gallety, G. D., Torispherical Shells—A Caution to Designers, *J. Eng. Industry,* Trans ASME, Feb 1969, 51–62.
4. Findlay, G. E., Moffat, D. G., and Stanley, P., Torispherical Drumheads: A Limit Pressure and Shakedown Investigation, *J. Strain Anal.*, 6(3) (1971), 147–166.
5. Bushnell, D., BOSOR-5 Program for the Buckling of Elastic-Plastic Shells of Revolution Including Large Deflections and Creep, *Computer and Struct.*, 6 (1976), 221–239.
6. Gallety, G. D., Design Equations for Preventing Buckling in Fabricated Torispherical Shells Subjected to Internal Pressure, *Proc. Inst. Mech. Eng.*, 200(A2) (1986), 127–139.
7. Kalnins, A. and Updike, D. P., New Design for Torispherical Heads (Welding Research Council Bulletin 364), June 1991, United Engineering Centre, New York.
8. Kalnins, A. and Updike, D. P., Elastic-Plastic Analysis of Shells of Revolution under Axisymmetric Loading (Welding Research Council Bulletin 364), June 1991, United Engineering Centre, New York.
9. Gallety, G. D. and Blachut, J., Buckling Design of Imperfect Welded Hemispherical Shells Subject to External Pressure, *Proc. Inst. Mech. Eng.*, 205C (1991), 175–188.

Local Loads and Supports

10. Wichman, K. R., Hopper, A. G., and Mershon, J. L., Local Stresses in Spherical and Cylindrical Shells Due to External Loadings (Welding Research Council Bulletin 107), 1965 (revised March 1979), United Engineering Centre, New York.
11. Bijlaard, P. P., Stresses from Radial Loads in Cylindrical Pressure Vessels, *Weld. J. Res. Suppl.*, 33 (1954), 615–623.
12. Duthie, G. and Tooth, A. S., Local Loads on Cylindrical Vessels; a Fourier Series Solution, in Behaviour of Thin Walled Structures, Rhodes and Spence, Eds., Elsevier Applied Science, 1984, 235–272.
13. Bijlaard, P. P., Additional Data on Stresses in Cylindrical Shells under Local Loading (Welding Research Council Bulletin 50), May 1959, United Engineering Centre, New York.
14. Bijlaard, P. P., Local Stress in Spherical Shells from Radial or Moment Loading, *Weld. J. Res. Suppl.*, 36(5) (1957), 240.
15. Bijlaard, P. P., On the Stresses from Local Loads on Spherical Pressure Vessels and Pressure Vessel Heads (Welding Research Council Bulletin 34), 1957, United Engineering Centre, New York.
16. Zick, L. P., Stresses in Large Horizontal Cylindrical Vessels Supported by Saddles Weld to the Vessel—A Comparison of Theory and Experiment, 3rd Int. Conf. Pressure Vessel Techn., ASME, New York, 1977, 25–38.
17. Tooth, A. S. and Nash, D. H., Stress Analysis and Fatigue Assessment of Twin Saddle Supported Vessels, Pressure Vessel and Components ASME Conf., ASME, New York, 1991, 41.
18. PD 9467, Stress in Horizontal Cylindrical Vessels Supported on Twin Saddle-Derivation of the Basic Equations Used in BS 5500, BSI, London, 1982.

19. Kupka, V., The Background to a New Design Proposal for Saddle Supported Vessels, *Int J. Pressure Vessel & Piping,* 46 (1991), 51–65.

Nozzles

20. Leckie, F. A. and Penny R. K., Stress Concentration Factors for the Stresses at Nozzle Intersections in Pressure Vessels (Welding Research Council Bulletin 90), 1963, United Engineering Centre, New York.
21. Eringen, A. C. et al., Stress Concentrations in Two Normally Intersections Cylindrical Shells Subjected to Internal Pressure (Welding Research Council Bulletin 139), 1969, United Engineering Centre, New York.
22. Rose, R. T., Stress Analysis of Nozzle in Thin Walled Cylindrical Pressure Vessels, *Brit. Welding J.,* 12(2) (1965).
23. Money, H. A., The Design of Flush Cylinder/Cylinder Intersections to Withstand Internal Pressure (CEGB Report R.D/B/N1061) CEGB, Berkley, Gloucester, UK, 1968.
24. Decock, J., Determination of Stress Concentration Factors and Fatigue Assessment of Flush and Extended Nozzles in Welted Pressure Vessels (Paper II-59), 2nd Int. Conf. Pressure Vessel Techn., San Antonio, Texas, Oct. 1973.
25. Decock, J., Reinforcement Method of Openings in Cylindrical Pressure Vessels Subject to Internal Pressure (CRIF Report MT104), Univ. Ghent, Belgium, 1975.
26. Mershon, J. L. et al., Local Stresses in Cylindrical Shells Due to External Loadings on Nozzles (Supplement to Welding Research Council Bulletin 107, WRCB No 297), 1965, United Engineering Centre, New York.
27. Moffat, D. G. and Mistry, J., Interaction of External Moment Loads and Internal Pressure on Variety of Branch Pipe Intersections, Proc. 6th Int. Conf. Pressure Vessel Techn., Pergamon Press, Oxford, UK, 533–549.

Exchanger Bellows

28. Standards of the Expansion Joint Manufacturers Association, EJMA, New York.
29. Kopp, S. and Sayre, M. F., Expansion Joints for Heat Exchangers, ASME Misc. Papers, Vol. 6, No. 211 (1950 ASME Annual Meeting).
30. Thomas, R. E., Validation of Bellows Design Criteria by Testing, ASME PV & Piping Conf., Orlando, Florida, 1982.
31. Smith, A. G., An Investigation into the Design of Thick Walled Bellows for Heat Exchangers, Ph.D. thesis, Paisley College of Techn., Paisley, Scotland, 1981.
32. Becht, C., Hong, C., and Skopp, G., Stress Analysis of Bellows, 11, ASME PVP-Vol. 51, 1981.
33. Stastny, R. J., Metallic Convoluted Expansion Joints, Application, Specification, and Installation, 61, ASME PVP-Vol. 51, 1981.

Fatigue

34. Langer, B. F., Design of Pressure Vessels for Low Cycle Fatigue, *J. Basic Eng.* (Trans ASME Series b), 84 (1962), 389–402.
35. Maddox, S. J., Fatigue Strength of Welded Structures, Abington Publishing, Abington, Cambridge, 1991.
36. Harrison, J. D. and Maddox, S. J., A Critical Examination of the Rules for the Design of Pressure Vessels Subject to Fatigue Loading, in Proceedings 4th Int. Conf. on Pressure Vessel Techn., Inst. Mech. Eng., London, 1980.
37. PD 6493, Guidance on Methods for Assessing in Acceptability of Flaws in Fusion Welded Structures, BSI, London, 1991.

LIST OF STANDARDS AND PROGRAMS

ANSYS	•	Finite Element Program from ANSYS Inc. 201 Johnson Road, Houston, PA
ASME III	•	ASME Boiler and Pressure Vessel Code, Section III, New York
ASME VIII	•	ASME Boiler and Pressure Vessel Code, Section VIII Division 1, New York
BS 5500	•	BS 5500: Specification for Unfired Fusion Welded Pressure Vessels British Standards Institution, London, UK
AD	•	Arbeitsgemeinschaft Druckbehälter (AD) Verband der Technischen Überwachungs-Vereine e.V. Postfach 103834, 45038 Essen, Germany
AD - B7	•	AD-Merkblatt B7 - Bolting
AD - B8	•	AD-Merkblatt B8 - Flanges
TRD	•	Technische Regeln für Dampfkessel (TRD) Verband der Technischen Überwachungs-Vereine e.V. Postfach 103834, 45038 Essen, Germany
TRD 301	•	Cylindrical Shells under Internal Pressure
TRD 309	•	Bolting
DIN 2505	•	Calculation of Flanged Joints Beuth Verlag GmbH, Burggrafenstraße 4-10, 1000 Berlin 30
KTA Code	•	Code of the Kerntechnische Ausschuss, Germany

LIST OF MATERIALS

DIN Norm Product Name	ASTM Norm
ST 35.8	A53 Gr.A
15 Mo 3	(A 161 Gr. T1)
W St E 255	A 442 Gr. 55 A 516 Gr. 55
W St E 355	– – – – – –
15 Mn Ni 63	– – – – – –
20 Mn Mo Ni 55	A 533
15 Ni Cu Mo Nb 5 (WB 36)	A 302 A 508 A 533
22 Ni Mo Cr 37	A 508
10 Cr Mo 910	A 387 A 542